The Easiest and Fastest Way To Manage Construction Projects of Any Types and Sizes

Construction Management
Step-by-Step Templates

- ➤ **5 Easy Steps** and **13 Success Keys** To Complete Your Project Successfully
- ➤ **Over 100** Templates, Flowcharts and Project Documents
- ➤ **Save Time** and **Money**
- ➤ For **Any Types** of Project

MARVIN M. GAMBOA, CE, PMP

Construction Management
Step-by-Step Templates

By: Marvin M. Gamboa, CE,PMP

www.ConstructionProjectManagementPro.com

CONSTRUCTION MANAGEMENT: STEP-BY-STEP TEMPLATES

Copyright © 2015 Pier Engineering and Consultants
All rights reserved.

No part of this Book may be reproduced or transmitted in any form or by any means, mechanical, electronic or by any information storage and retrieval system, without a written permission from the author or publisher.

ISBN: 978-971-95901-8-7

Pier Engineering and Consultants

Project Managers • Consulting Engineers

#0754, Zone 5, Maliwalo
Tarlac City, Philippines
+63(045) 491-4994
www.PierEngineeringandConsultants.com
info@pierengineeringandconsultants.com

Legal information and disclaimer: All the contents and information contained here are from author's knowledge, trainings and experiences, the purchaser of this publication assumes full responsibility for the use of this Book. This publication does not constitute legal and financial advice and the example project, format, document and data here are not intended to represent or guarantee that everyone in construction project management or construction management will achieve same results.

The condition of project will vary from region to region and from one country to another, it is highly recommended that the purchaser of this publication will consult professional, legal and professional authorities based on their project geographical area. Any legal templates and document here must be reviewed by legal professionals and authorities in your project location.

Contents

INTRODUCTION		1
STEP 1	Initiating	4
STEP 2	Planning	18
STEP 3	Executing	90
STEP 4	Monitoring and Controlling	142
STEP 5	Closing	179
TRAINING	Manage Templates and Documents the Easy Way	188
RESOURCES		193

Introduction

Welcome!

Welcome to Construction Management: Step-by-Step Templates. In this Book we are going to talk about the 5 steps and 13 success keys to manage construction projects, one of the main reasons we need to explore about the construction management or the construction project management is because until now most of the construction projects are resulted to failed projects, it is roughly 70 to 85 percent of the construction projects are failed.

The purpose of this publication is to present the applicable group of processes, procedures and techniques from project management standards for construction project in a simplified and direct approach.

I am a registered Civil Engineer (CE) and a Certified Project Manager (PMP). I have been in the construction industry for more than 15 years now, from junior to senior engineer up to project management level, I have seen project managers that failed and succeeded for their construction projects. It takes the right knowledge, experience and right tools to achieve success in project management for construction project, this Book will guide and help you succeed in any construction projects you handle.

I have numerous free templates, articles, reports, guides and tips for construction management or construction project management to help professionals or any stakeholders within construction industry. You can access free templates, information and documents in my blog Construction Project Management Pro , you can bookmark it and RSS feed email to keep you updated for all free templates, documents and other resources for your construction projects.

Thank you for purchasing this Book and if you have any questions, comments and feedback , good or bad , I would love to hear from you, please let me know at send me info .

I hope it would be very helpful and find it valuable.

To Your Success in Managing Construction Projects

Marvin M. Gamboa

Who can use Construction Management: Step-by-Step Templates?

The primary audience to this publication is the project manager for construction projects, to help them manage the project from start to hand over and from simple to complex construction projects .And this publication can also be used but not limited to Construction Manager, Engineers, Architects, Project Owner, General Contractor, Contractor and Sub-contractor, Supplier and Manufacturer, Construction Authorities, Students, Construction related professionals and any other stakeholders for construction industry.

Project Management versus Construction Management

Project Management is the use of knowledge, skills, tools and processes to deliver the specified project requirements, most of the time construction management has been separated to project management , though construction management is now a profession or some countries offer this as a course , as a project management professional , construction management is under the umbrella of project management where in, you are managing a construction project a very special project under project management but still managing construction project or construction management is a project management for construction project, other industries you can apply project management are I.T, Pharmaceutical etc.

In this Book, the consideration is , project management for construction project is the same as construction management or construction project management *(this is generally the managing of a construction project).* Project management as a whole is composed of 9 keys (*knowledge areas)* while the Project management for construction projects have 9 keys plus construction extension (additional process groups) and namely, safety and health, environmental, financial and claims management, in short there 13 keys (*Knowledge areas*) covering the project management for construction.

Step 1
Initiating

www.ConstructionProjectManagementPro.com

STEP 1 : INITIATING

Initiating is the first step in managing construction projects. It is composed of processes aimed to authorize the project, identify specific phases, determine stakeholders, assign project manager(s), and formally start the construction work.

Flow Chart :

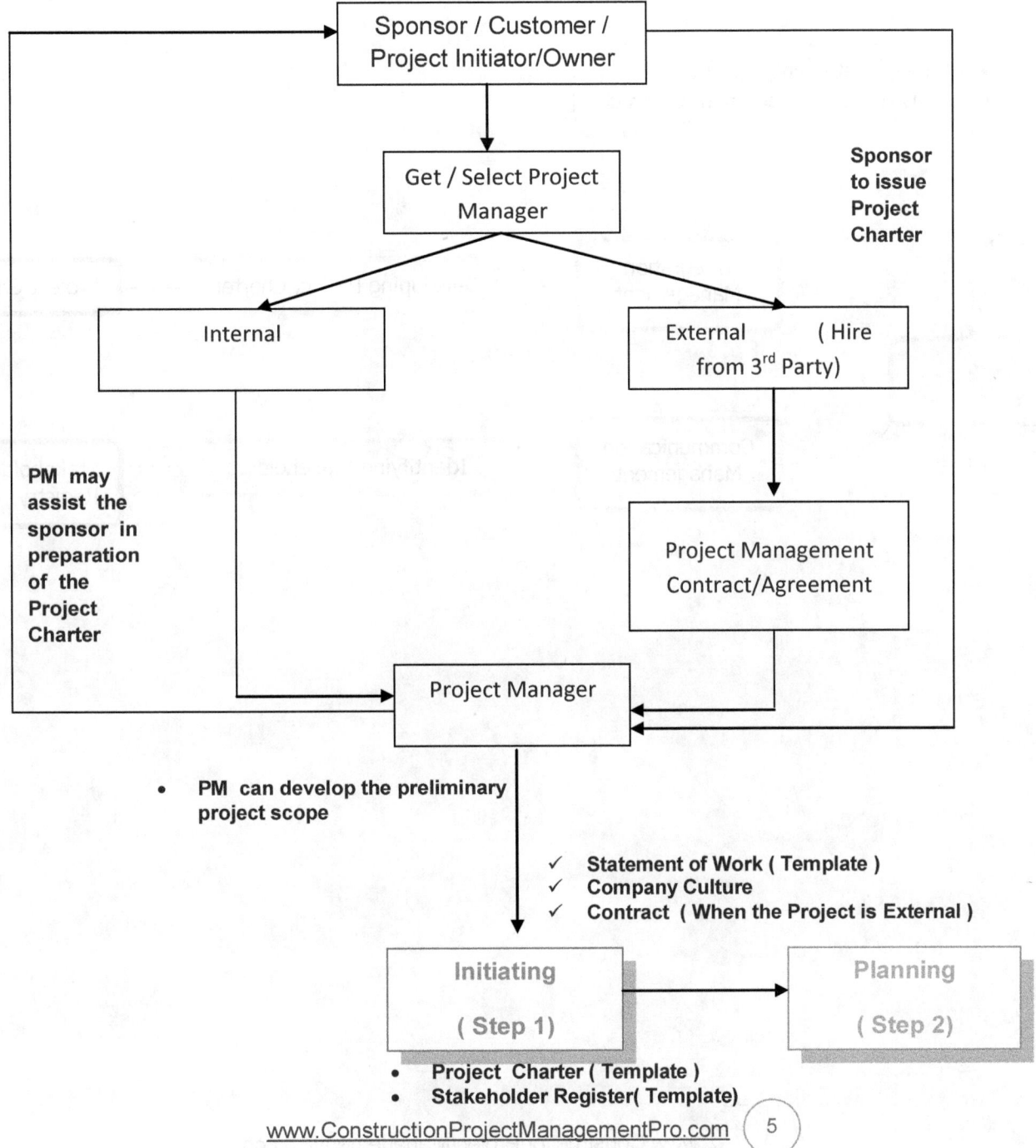

Mapping:

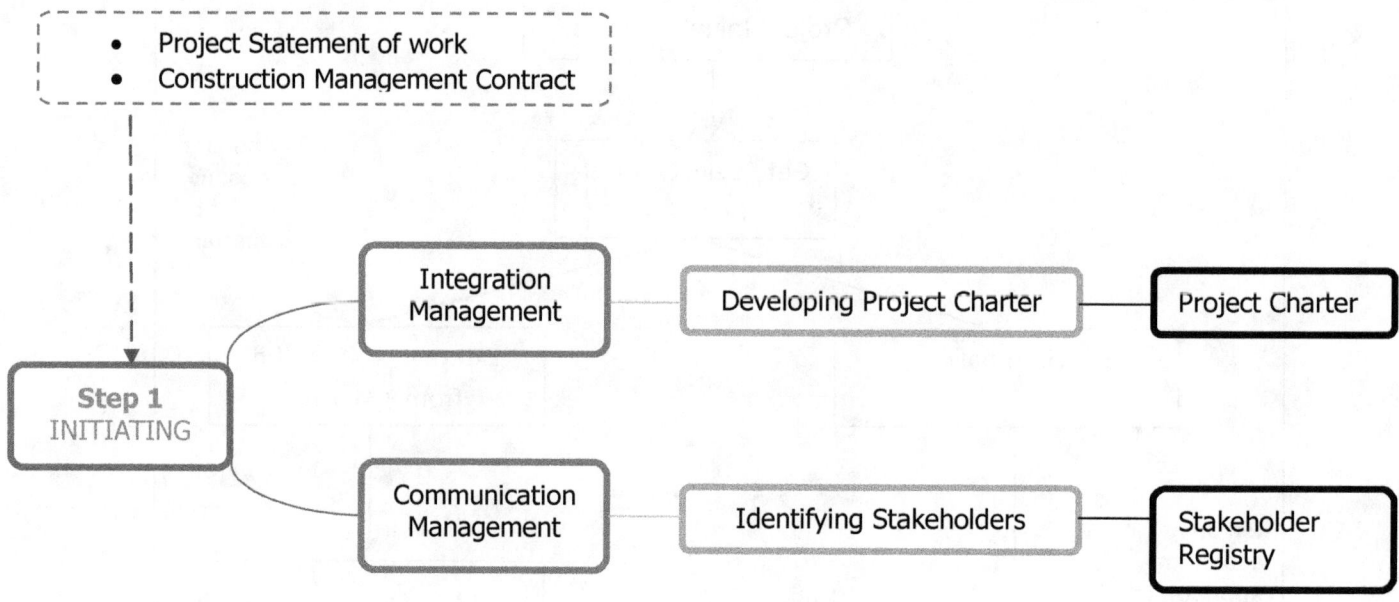

Construction Management Contract

The Construction Management Contract is a legal document subject for remedy in court. It serves as a legal binding agreement between two parties: (1) the sponsor or customer, and (2) the construction project manager or service provider. It can also be a contract between sellers and buyers.

Purchase Order, more commonly known as P.O., is a simple form of contract.

Template : Construction Management Contract

Project Statement of Work

For internal projects, the Statement of Work is a document to be created by the customer or sponsor describing the scope of the entire project, services to be rendered, and needs to be delivered.

For external projects, the Statement of Work can be a part of bid documents, such as request for bid, or part of the contract from the customer. The document will serve as an initial overview of the project and an input needed in the preparation of project charter.

Project Statement of Work may not be complete or comprehensive, as project details will be further defined in the project scope statement.

Template : Project Statement of Work

PROJECT CHARTER

The Project Charter recognizes and authorizes the project. It formally appoints or assigns a project manager to a specific project and defines his level of authority. It is being created in the construction management phase.

In a construction project, some sponsors or customers initially hire architectural designers who will allow them to see the overall outcome of the project before getting a project manager. On the other hand, there are sponsors or customers who look for project managers just to manage the actual construction project.

This is risky. The project manager should be involved in the initial stage of the project, particularly during the design stage. When the project charter has been issued, the project manager will now get the team.

- The project will not exist without the project charter.

- The project charter formally authorizes the project.

- This will be issued by the sponsor. However, the project manager can assist the sponsor in preparing the document.

Template : Project Charter

STAKEHOLDER REGISTRY

Provides information on the interest of the stakeholder. This document is ideally used to identify the stakeholder who can supply all the data regarding the project requirements. Generally, this document is created during the first step of the process which is the construction management phase.

Template : Stakeholder registry

Reference No. : CM-PEC-D-0001-2015

Template ID : IT-I-003
Revision : 000

Date : February 28 , 2015

PROJECT CHARTER

1.0 PROJECT DESCRIPTION

This shows a descriptive summary of the project.

The Management of P.E.C. School have noticed the increase of college students every school year and decided to have an additional school building to accommodate the increasing number of enrollees on each offered courses. The additional school building will be constructed adjacent to building A inside the school campus. The proposed school building will consist of eight (8) floors and will included a basement and a roof deck, with an average floor area of 2,000 square meters per level.

2.0 PROJECT PURPOSE

This enumerates primary reason(s) why the project is needed.

The project aims to construct a new school building for the increasing number of students who will enroll for our existing and new courses. The new building to be constructed will include an auditorium, cafeteria, computer laboratory, case rooms, computer rooms, classrooms, offices, landscape, ramps, pathways, and other areas necessary for learning.

3.0 PROJECT JUSTIFICATION

This shows a brief explanation of the project in terms of business.

This project is to be completed before the opening of the school year 2013. We expect the increase of enrollees for our existing and new courses to be offered; it will attract new students in our region and nearby areas, and thus, will result to increase in revenue for our school by at least $600,000 for the year 2013.

4.0 PROJECT MANAGER

This enumerates the people will lead and manage the project.

Name	Organization	Contact Details	Assignment Date	Authority Level
Pier John	Pier Engineering and Consultants	Tel: Email:	April 16 , 2013	• Manage and Lead the Project • Manage and Approve Cost baseline • Manage and Approve Schedule baseline

www.ConstructionProjectManagementPro.com

				- Determine and Approve final Project Budget - Determine and Approve Team Members - Manage and Approve Changes

5.0 RESOURCES

This identifies who and what resources will be provided to finish the project.

Name	Role	Organization	Contact Details	Assignment Date	Remarks
Marvin Gamboa	Project Leader	Pier Engineering and Consultants	Tel: Email:	May 16, 2013	Assigned
Name	Contractor's Leader	Company X	Tel: Email:	May 16, 2013	Assigned
Others					To be determined by the Project Manager

6.0 STAKEHOLDERS

This identifies the group of people who will be the affected, have impact on, and benefit from the project.

Name	Role / Interest
Construction Team	The construction team will be the responsible in delivering all necessary requirements for the successful completion of the project.
P.E.C Management	The President of P.E.C School is the sponsor of the project and should be informed about the updated project status.
Pier Engineering and Consultants Team	Pier Engineering and Consultants Team will lead the entire project until its completion.
Local Government and Authority	Local government agencies and authorities shall implement the standards and area ordinances and ensure that these will be observed throughout the project.

Reference No. : CM-PEC-D-0001-2015　　　　　　　　　　Date : February 28 , 2015

Designer's Team	The designer's team are the ones responsible in ensuring that the established standards, plans, and specifications will be properly applied throughout the project.
Campus Faculty, Students, Groups, and Public.	The campus faculty, together with the students, organizations, and even the public, will also ensure that the established standards, regulations, and all requirements are applied throughout the execution of the project

7.0 SCOPE

This identifies the boundaries, limitations, and range of the portion of the project.

The scope will cover the following items:

01. Civil Works
02. Architectural Works
03. Site/Earth Works
04. Concrete Works
05. Metal Works
06. Masonry Works
07. Moisture and Thermal Protection
08. Curtain Walls, Louvers, Doors, and Windows
09. Sky Light and Canopy
10. Finishes
11. Plumbing Works

The items below, on the other hand, are not included in the scope initially. P.E.C School will be the one responsible in providing these items:

01. Electrical Works

    - Lighting and Power System
    - Fire Alarm and Detection System
    - Auxiliaries
        - Electronic Safety and Security Works
        - Audio / Video System
        - Public Address (P.A. System)
        - Structured Cabling Works
        - Telephone Wiring System
    - Supply and Installation of Electric Generating Set
    - Lightning Protection System

02. Mechanical Works
    - Heat ,Ventilation and Air conditioning System (HVAC)
    - Supply and Installation Elevator Equipment

03. Supply of Lighting Fixtures

Reference No. : CM-PEC-D-0001-2015   Date : February 28, 2015

04. Supply of Plumbing Fixtures
05. Supply of Pumps and Pump Controllers
06. Supply of Air conditioning Equipment
07. Supply Floor / Wall Tiles and Carpets
08. Supply and Installation Operable Wall System
09. Supply Furniture /Chairs / Loose Cabinet / Glass Whiteboard
10. Supply of Fire Extinguishers
11. Supply of Signage and Directories
12. Supply Sewage Treatment Plant
13. Supply of Dry Type Transformer For Elevator (when necessary)
14. Supply Fans / Blowers and Exhaust Fans.

8.0 DELIVERABLES

This specifies the expected results of the project.

Categories	Description
Building Structure	A complete and fully-furnished eight-storey building as part of the P.E.C School, including the roof deck for auditorium use and the basement as canteen and offices
Services	Functional and operational supply of electricity, water, lightings, air conditioning, telephone lines, internet, fans and blowers, sewage treatment, drainage, fire alarm and detection, lightning protection and elevators
Building Access	Ramps, pathways, and stairs
Facilities	Complete, fully-furnished, and functional classrooms, case rooms, faculty rooms, computer laboratory rooms, offices, auditorium, canteen, and coffee shop.

9.0 CONSTRAINTS

This specifies the established limits and/or restrictions of the project

01. The new P.E.C. School Building will be finished on or before April 16, 2014.
02. The allotted budget for the project is $10,000,000.
03. The allowed time for concrete pouring is from 6:00 PM to 7:00 AM only, when classes are not being held in the school property.
04. The delivery of project materials inside the school vicinity from 6:00 PM to 7:00 AM only.
05. The project site will maintain the "no noise and no dust" policy.

Reference No. : CM-PEC-D-0001-2015

10.0 ASSUMPTION

These are additional statements related to the project that are believed to be true even without proof.

- The Design and Technical Team will be available on site to assist the construction activity.
- All the required resources and services are available in the area and nearby provinces.

Approval:

Pier Engineering and Consultants

By:

_____

Pier John

Project Manager

P.E.C School

By:

_____ _____

Alexander Thomas   **Richard Ramos**

Vice President   President

Stakeholder Registry

Company Address :
Telephone :
Fax :
Email : info@
Website : www...

Date Prepared : 02/28/15	**Template ID:** IT-COM-001	
Revision : 000	**Pages :** 1 of 1	
Reference No. : CM-PEC-D-0002-2015		

Project	Name of the Project
Project Package	Name of the package / phase of the project
Project Location	Name of the location of the project

Name	Position (Company)	Role (Project)	Contact Details	Requirements	Expectation	Influence	Category	Power		Interest		Analysis	
								(A) Low	(B) High	(C) Low	(D) High	Impact Assessment	Possible Strategies (to gain support / minimizing hindrance)
Pier John	Senior Engineer	Project Manager	Mobile: Tel: email				Internal/External/resistor/supporters etc.						

When :
B-C (keep contented)
B-D (Manage carefully)
A-C (Monitor)
A-D (continuously Informed)

Template ID : IT-I-002
Revision : 000

Reference No. : CM-PEC-D-0003-2015

Date : February 28, 2015

PROJECT STATEMENT OF WORK

1.0 PROJECT DESCRIPTION
This provides a descriptive summary of the project.

The proposed school building will have eight (8) floors, including basement and roof deck, and will measure an average floor area of 2,000 square meters per level or floor. The said building will be constructed inside the school campus and will include facilities such as coffee shop, canteen, computer laboratory, case rooms, computer rooms, classrooms, offices, landscape, ramps, pathways, and an auditorium.

The primary purpose of this project is to accommodate new enrollees for the existing and new courses being offered in the school.

2.0 PROJECT SCOPE DESCRIPTION

- Civil Works
- Architectural Works
- Site/Earth Works
- Concrete Works
- Metal Works
- Masonry Works
- Moisture and Thermal Protection
- Curtain Walls, Louvers, Doors, and Windows
- Sky Light and Canopy
- Finishes
- Plumbing Works

Not covered in the scope:

- Electrical Works
  Lighting and Power System
  Fire Alarm and Detection System
  Auxiliaries
  Supply and Installation of Electric Generating Set
  Lightning Protection System

- Mechanical Works
  Heat, Ventilation and Air conditioning System (HVAC)
  Supply and Installation Elevator Equipment

Reference No. : CM-PEC-D-0003-2015	Date :	February 28 , 2015

- Supply of Lighting Fixtures
- Supply of Plumbing Fixtures
- Supply of Pumps and Pump Controllers
- Supply of Air conditioning Equipment
- Supply Floor / Wall Tiles and Carpets
- Supply and Installation Operable Wall System
- Supply Furniture /Chairs / Loose Cabinet / Glass Whiteboard
- Supply of Fire Extinguishers
- Supply of Signage and Directories
- Supply Sewage Treatment Plant
- Supply of Dry Type Transformer For Elevator (when necessary)
- Supply Fans / Blowers and Exhaust Fans.

Prepared by: _____

P.E.C School

_____   _____

Alexander Thomas   **Richard Ramos**

Vice President   President

www.ConstructionProjectManagementPro.com

Reference No. : *(Refer to your legal Advisor/ Attorney)*	Template ID : IT-I-001 Revision : 000 Date : September 01 , 2013

CONSTRUCTION MANAGEMENT SERVICES AGREEMENT

BELOW ARE THE COMPONENTS OF THE CONSTRUCTION MANAGEMENT CONTRACT BUT NOT LIMITED TO; *(Please refer to your legal advisor)*

1. ARTICLE I
   - DEFINITION OF TERMS
2. ARTICLE II
   - THE SERVICES
3. ARTICLE III
   - OBLIGATIONS OF THE OWNER
4. ARTICLE IV
   - PERSONNEL
5. ARTICLE V
   - CONTRACT AMOUNT AND METHOD OF PAYMENT
6. ARTICLE VI
   - INDEMNITY
7. ARTICLE VII
   - FORCE MAJEURE AND OTHER CAUSES
8. ARTICLE VIII
   - MODIFICATION
9. ARTICLE IX
   - TERMINATION OF SERVICES
10. ARTICLE X
    - LIABILITY OF THE CONSULTANT
11. ARTICLE XI
    - DISPUTE AND ARBITRATION
12. ARTICLE XII
    - OWNERSHIP OF STUDIES, DATA AND PLANS
13. ARTICLE XIII
    - COMMENCEMENT AND DURATION OF CONTRACT
14. ARTICLE XIV

Step 2
Planning

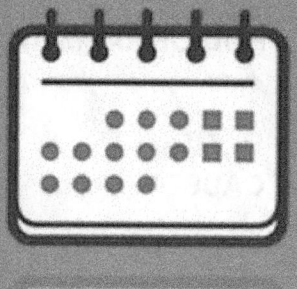

www.ConstructionProjectManagementPro.com

STEP 2 : PLANNING

Planning is the second step in managing construction projects. It is composed of process groups aimed to establish a concrete project management plan and identify all the necessary documents for the project. These process groups enable the stakeholders and project managers to create a comprehensive and detailed scope, schedule, costing, and other required information that will help them manage and control established plans for the project.

Flow Chart : Project Management process is not always sequential or performed in identical

```
Initiating (Step 1)  ──────►  Planning (Step 2)
                                      │
 • Step 1 is already                   ▼
   completed.              Creation of project
                            management plan
                                      │
                                      ▼
   ✓ Approved             ┌─────► Scope ◄───── ✓ Project team has
     Changes,             │          │              been selected
     corrective and       │          ▼              and created in
     preventive action    │        Time             this stage
                          │          │
                          │          ▼
                          │        Cost
                          │          │
                          │          ▼
                          │       Quality
                          │          │
                          │          ▼
                          │       Human
                          │      Resource
                          │          │
                          │          ▼
                          │    Communications
                          │          │          ✓ When risks
                          │          ▼             triggered
                          └─────► Risks ◄────      changes,
                                     │             necessary
                                     ▼             adjustment
                                Procurement        is to be
                                     │             applied
                                     ▼
                              Final and
                              Approved
                              project       ──►  Executing (Step 3)
                              management
                              plan
```

Construction Management: Step-by-Step Templates — Step 2 : Planning

Mapping:

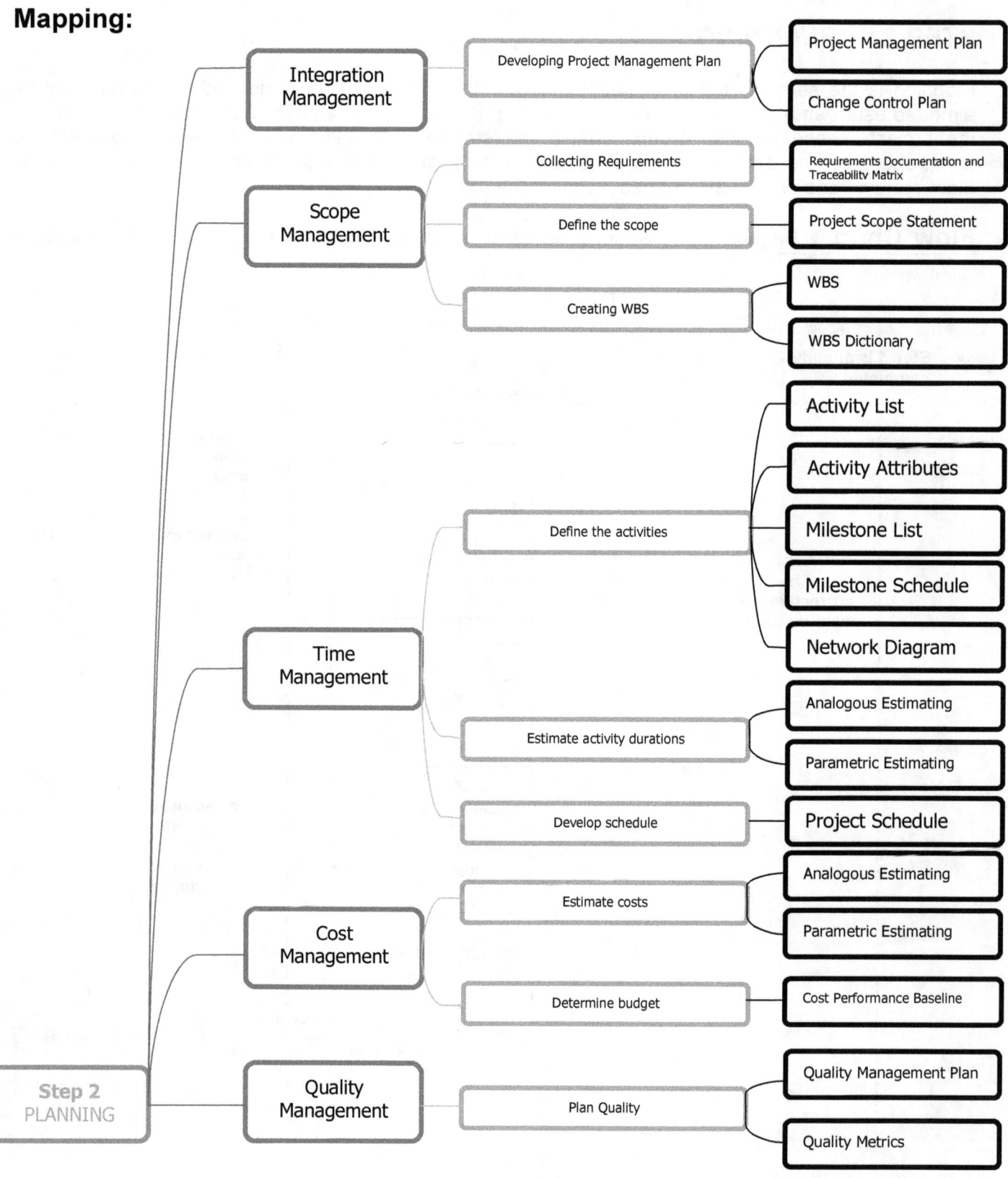

www.ConstructionProjectManagementPro.com

Step 2: Planning

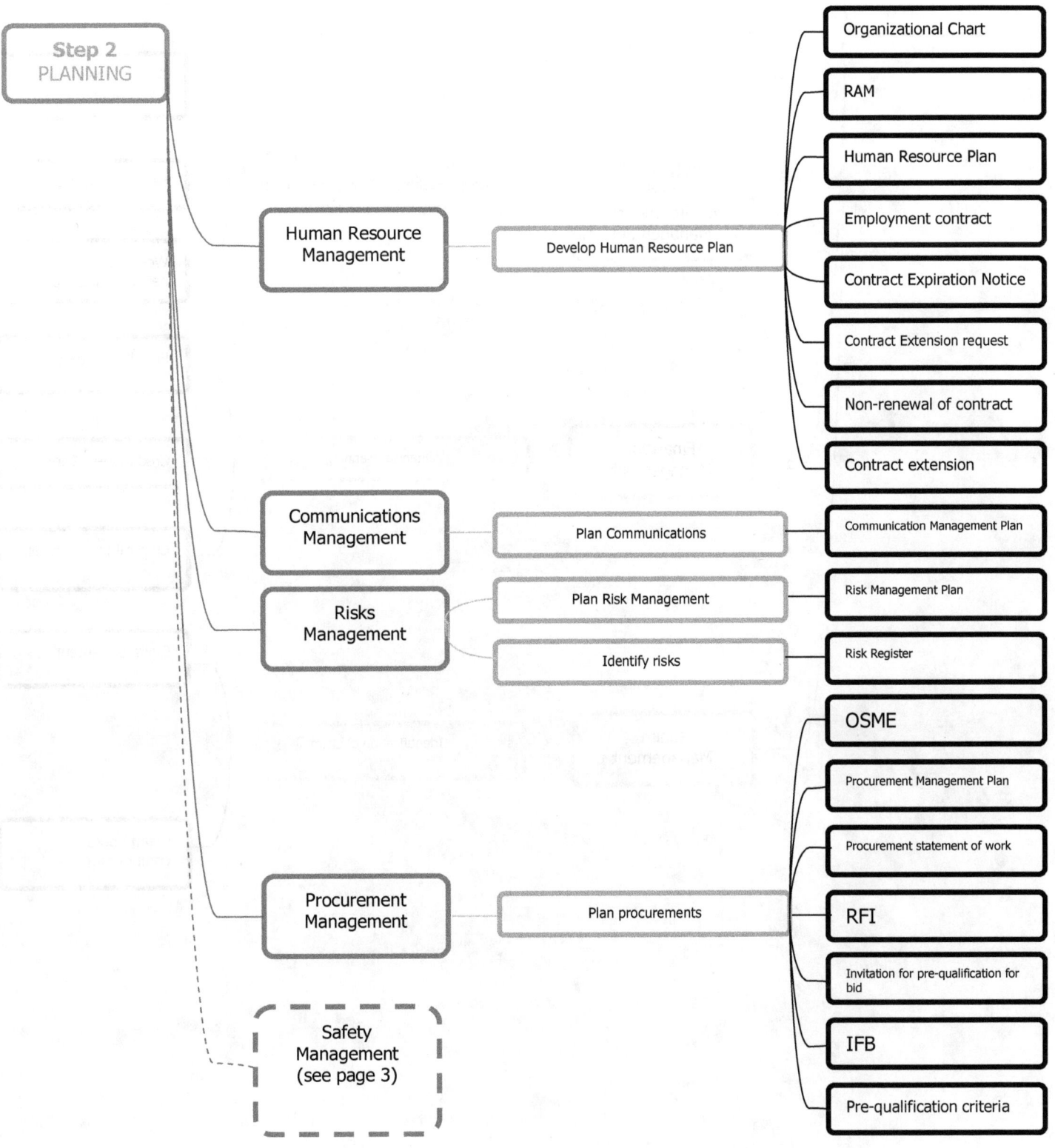

Construction Management: Step-by-Step Templates — Step 2 : Planning

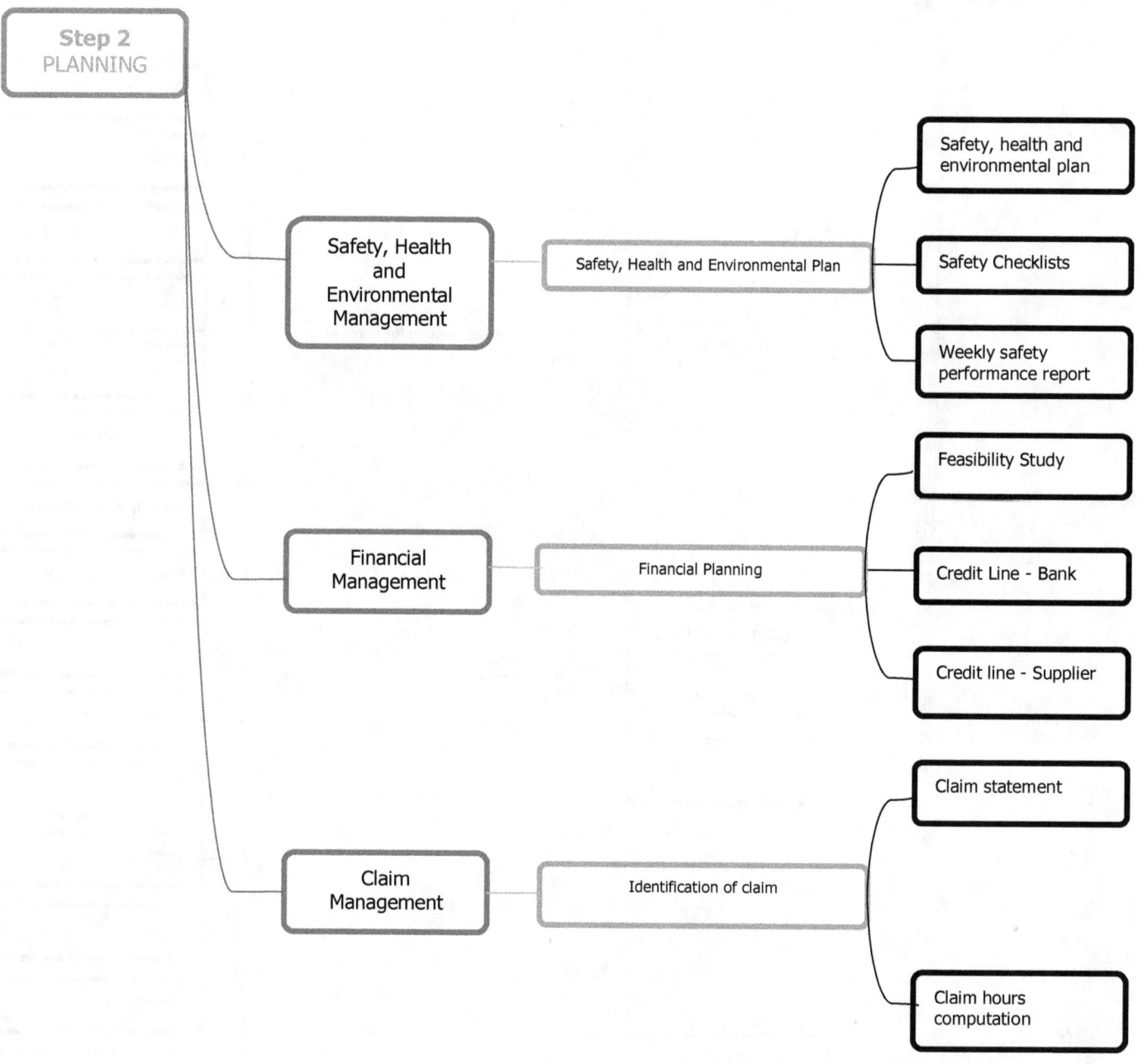

PROJECT MANAGEMENT PLAN

Is an approved plan that describes how the project will be executed, monitored, controlled, communicated, and closed out. This document follows a specific type of format either summary type or a more detailed outline. It also includes some subsidiary management plans listed below.

Template : Project Management Plan

CHANGE CONTROL PLAN

Is a vital part of the project management plan, which consists of the processes of analyzing, evaluating, approving, and managing changes to project document, deliverables, and the project management plan itself. It is often being done from the start to completion of the project.

Template : Change Control Plan

REQUIREMENTS DOCUMENTATION AND TRACEABILITY MATRIX

Requirements documentation is a list of all requirements associated with the project and can be categorize as stakeholder and priority, or trade and technical. Lists of requirements will be progressively elaborated as the data becomes available on the project. This document is usually created during the second step of managing construction projects.

Also included in this form is the Requirements Traceability Matrix which is used to track the given requirement, link to a project purpose, and verify the identified requirements.

Template : Requirements Documentation and Traceability Matrix

---

Hot Tip

Document the sponsor's or customer's requirements in terms of the material supply or specific trade that they can provide. This must be noted in the early stage of planning, which is often called Owner Supplied Materials (OSM) because some owners or sponsors have their related businesses in the material and equipment supplies.

When customers or sponsors have their repeated projects like warehouses or shopping malls, they prefer to select some materials or equipment that they can use for their next project. They can remove the items with mark ups (upcoming contractor), so that they can save money from it.

PROJECT SCOPE STATEMENT

Is where the project deliverables are described in details and the necessary requirements, including expected work and output, are elaborated .This serves as an important guide for the execution of the project by the team members and provide the team with the detailed planning approach.

Furthermore, project scope statement provides in-depth information for change request as additive or deductive. This document is often created during the second step of managing construction projects.

Template : Project Scope Statement

Work Breakdown Structure(WBS)

Is where the project will be broken down into smaller manageable pieces. It is designed to be in a top-down decomposition of the project with the lowest level as the work package. The representation of work breakdown structure involves a hierarchical chart format and breakdown lists format. This document is generally created during the second step of managing construction projects.

Key Point

- The top level is the project title.

- The project must be broken down into smaller and manageable pieces, all the way up to work package level. There is no rule of thumb in breaking down the project into WBS. It can be outsourced or contracted, can be estimated, cannot be logically divided, and deliverables are to be defined.

- Provide numbering for easy identification and further use, such as cost coding.

- WBS should be created with the help of the team.

Template : WBS

WBS DICTIONARY

Work breakdown structure dictionary provides supplemental details to WBS, such as detailed description of work to be done, detailed information of work packages, and control accounts for cost. This document is often created during the second step of managing construction projects.

Template : WBS Dictionary

Hot Tip

- Work not included in the WBS is not part of the project.
- You cannot finish a successful project without WBS. This is the foundation of any project.
- With every project you have, document the WBS and use the previous one as template for future similar projects.

ACTIVITY LIST AND ATTRIBUTES

Is a list of all activities associated with the work packages to complete the project. The work package from our WBS is to be decomposed further in to activity level, where it can easily managed, monitored, and estimated. Our lists of activity will then be arranged and sequenced for the next process.

Together with the form is the Activity Attributes. We establish the more detailed description of an activity by specifying the resources needed for this activity assumption and constraints and predecessor and successor activity.

These documents are to be created during the second step of managing construction projects.

Templates : Activity List
Activity Attributes

MILESTONE LIST

Specifies the relative time or event for the project. The document comprises all the milestones – whether optional, mandatory, or based on historical data and information. This document is often created during the second step of managing construction projects.

Template : Milestone list

NETWORK DIAGRAM

Is a type of diagram where activities involved in the project are to be arranged and sequenced according to how the entire work will be performed as a whole group of activities. It shows the dependencies of activities using critical path or time scaled diagram. This document is created during the second step of managing construction projects.

There are two (2) ways on how to draw network diagram. First is the Precedence Diagramming Method (PDM) and the second one is the Arrow Diagramming Method (ADM). The PDM or Activity-on-Node (AON) involves boxes or nodes that represent activities connected by arrow that symbolize their dependencies. On the other hand, ADM or Activity-on-Arrow (AOA), involves arrows that represent activities and boxes that symbolize dependencies. This type of diagram usually has finish-to-start relationships and may use dummies for activities.

Template : Network Diagram

ACTIVITY DURATION ESTIMATES

Refers to the estimated working period that will take to finish the established activity. Duration estimates can be performed by using analogous estimating, parametric estimating, or three-point estimating. This document is usually created during the second step of managing construction projects.

Template : Analogous Estimating
Parametric Estimating

PROJECT SCHEDULE

Is a representation of a desired format where the activities of the project show the starting date, end date, completed activities, milestones, and relationships. It can be presented in various formats, including tabular, bar charts, Gantt charts, milestone charts, or schedule network diagrams. This document is often created during the second step of managing construction projects.

- Unrealistic schedule is project manager's fault!

Template : Project Schedule

Construction Management: Step-by-Step Templates
Step 2 : Planning

ACTIVITY COST ESTIMATES

Is the application of measurable method to establish costs of the resources needed to complete the activity. Cost estimates are often expressed in currency units, such as dollar and peso, and will be progressively elaborated and refined when data and information are available.

Generally, cost of an activity refers to costs pertaining to materials, manpower or labor, equipment, services, and contingencies. Activity cost estimates can be performed by using analogous estimating, parametric estimating, three-point estimating, or bottom-up estimating.

This document is to be created during the second step of managing construction projects.

Templates: <u>Analogous Estimating</u>
<u>Parametric Estimating</u>

COST PERFORMANCE BASELINE

Is a time phased budget at completion of the project. This is commonly presented in the form of S-Curve for monitoring and controlling the cost for the project. This document is often created during the second step of managing construction projects.
Templates: <u>S-Curve</u>

QUALITY MANAGEMENT PLAN

Is part of or a subsidiary plan of the project management plan. This plan includes establishing methods, processes, and procedures to be able to meet the defined requirements of the project. It is important to establish what the acceptable deliverables are, what to do to meet the requirements of being acceptable, and how it will be measured against the requirements. This document is being created during the second step of managing construction projects.

- For team members, always check the quality by self-inspection. The entire organization has responsibilities in terms of quality.

- Poor quality of the project is project manager's fault!

- Spend more time in improving the quality and processes.

- Spend more time and money in quality concerns..

Template: <u>Quality Management Plan</u>

QUALITY METRICS

Includes the exact values that are to be measured on a certain project. This includes the acceptable measurement, including limits and controls in terms of specific values. The values in the metrics are to be measured during the quality control process , and knowing the limits from the metrics will result to acceptable deliverables and corrective or preventive actions. This document is being created during the second step of managing construction projects.

Step 2 : Planning

Construction Management: Step-by-Step Templates

> You can also check approved drawing plans, architectural, structural, and all required trades. All measurements, including vertical height, horizontal, thickness, inclined, elevations, diameters, and gauges, are parts of quality metrics with direct measurement approach. It is good to bring the drawing on site and measure actual sizes as per approved drawing plan (signed and marked with FOR CONSTRUCTION). Other bring smaller printed version (like A3 size) on site for convenience with blow up details. *(See sample template)*
>
> Tests like compressive strength of concrete and tensile for reinforcing steel can be conducted. However, take note that when testing is being done, be sure to have an inspector or technical consultant with you to check and verify the calibration of the testing apparatus or equipment being used. *(See sample template)*

Templates: Quality Metrics

ORGANIZATIONAL CHART

Is a document that highlights the role and responsibility of the team and each person comprising it. The chart also features each work package under work breakdown structure that has a definite owner and clear understanding of their respective roles and responsibilities. It also serves as a graphic representation of relationships and positions in a top-down format. This document is being created during the second step of managing construction projects.

> **Hot Tip**
> - Create and post organizational chart inside the office. You can create one per team or per group, so that contractors, for example, have their own chart in their office.
> - You can also post an organizational chart in the conference room, and include basic contact and picture of the whole team with their key persons like contractor, project management team, and other consultants for quick reference.

Templates: Organizational Chart

RESPONSIBILITY ASSIGN MATRIX (RAM)

Responsibility Assign Matrix or RAM is a document that shows the relations between work packages or project activities and the project team. Through this document, all the tasks and activities concerning a specific person or team can be defined and established, serving as a comprehensive guide to each team members involved in the project. RAM uses the RACI method – Responsible, Accountable, Consult, and Inform. This document is being created during the second step of managing construction projects.

Template: RAM

HUMAN RESOURCE PLAN

Is an essential component of the project management plan that creates a system on how the human resources will be staffed, managed, and released. This document is being created during the second step of managing construction projects.

Templates: Human Resource Plan, Employment Contract, Contract Expiration Notice, Contract Extension Request, Non-Renewal of Contract, Contract Extension

COMMUNICATION MANAGEMENT PLAN

Is another essential component of the project management plan. It involves a system on how to address information for stakeholder that answers what form of communication, when, and how frequent should it be done. This document is often created during the second step of managing construction projects.

Template: Communication Management Plan

RISK MANAGEMENT PLAN

Is another essential component of the project management plan, designed to deal with possible risks of the project and preventing issues that may result to negative or bad effect. In return, it specifies steps on how to come up with positive results or good effects by establishing plans in the first place before any problems might occur. This document is to be created during the second step of managing construction projects.

Template: Risk Management Plan

RISK REGISTER

Is a list of identified risks that may occur during the project execution. The list can be identified by using some tools and techniques, including checklist analysis that evaluates risks from the lowest level of risk breakdown structure to the highest. This document is usually created during the second step of managing construction projects.

Template: Risk Register

PROCUREMENT MANAGEMENT PLAN

Is another essential component of project management plan that addresses the procurement process, including documentation, managing, and closure of contracts. This document is often created during the second step of managing construction projects.

Templates: Procurement Management Plan, Owner Supplied Materials and Equipment (OSME)

PROCUREMENT STATEMENT OF WORK

Is a document that can be developed from the project scope. It comes from a portion of the project scope to be considered for contracting packages. This should include detailed information such as location of work, quantity, specification, specified quality, and work schedule. This document is often created during the second step of managing construction projects.

Template: Procurement Statement of Work

PROCUREMENT DOCUMENTS

Are written materials that ask for proposals from prospective sellers. There are different types of procurement documents, including request for information (RFI), request for proposal (RFP), request for quotation (RFQ), invitation for bid (IFB), and tender notice. This document is often created during the second step of managing construction projects.

Templates: Request For Information (RFI), Invitation For Prequalification, Invitation For Bid (IFB)

SOURCE SELECTION CRITERIA

Is a type of document that is often included in the procurement documents. We establish criteria to rate the sellers' proposals in terms of points or score. Selection criteria may be broad or simple, depending on the extent of the project. Criteria are usually based on financial proposals and technical proposals.

This document is being created during the second step of managing construction projects.

Template: Prequalification Criteria

EHS PLAN (Environmental, Health and Safety Plan)

Project Safety Management is one of the extensions considered in project management for construction project, safety is usually combined with health management. Safety management processes are activities of the team and the project owner to determine safety policies and requirements to prevent injuries, accidents, fatalities and damage to properties.

Project Environmental Management is another extension considered in project management for construction project .Environmental management processes are activities of the team and the project owner to determine environmental policies and requirements to prevent possible impact of the construction project to environment, to promote and achieve environment conservation and potential nature improvement.

Now a days, construction companies and government agencies have deeply concern about Health , Safety and Environmental factors . Other companies have sets of standards on these and they commonly combined in one plan , like EHS Plan(Environmental, Health and Safety Plan) , every construction projects you will handle with different location or geographical area will have a unique considerations in terms of safety, health and environmental , depending with the complexity and nature of the project . I've attached the basic health and safety plan/manual here, a short version for you to familiarize on what is to be stated and needed.

Construction Management: Step-by-Step Templates

Step 2 : Planning

Hot Tip

Prior to selection of General Contractor, they must submit Safety, Health and Environmental Plan.

Always ask for safety practitioner. For equipment operators, see to it they are legitimate/certified (like tower crane operators must be certified or accredited by your project geographical location) , operators also must passed the drug test in the country of your project.

Check the requirement on generators, other Countries requires the permit to operate as requirement for environmental factor. Prior to installation, contractor must satisfy the government agency requirements.

Key Point

- The project will not be started without safety personnel, some countries they require safety practitioner.

- Some Countries require you to submit EHS Plan for their approval.

- Check geographical standards and requirements for your construction projects.

Templates: Safety and Health Plan, Safety Checklist, Weekly Safety Performance

FINANCIAL MANAGEMENT

In this publication, processes that includes project funding is more on financial resources and management of funds, financial management here is not the cost management where in the cost management requires the daily management of the labour , material and equipment cost. Different types of financing the construction projects is to be used, some are but not limited to, Build-Own-Operate (BOO),Build-Operate-Maintain (BOM), Joint Venture ,privatization, etc.

This Publication focus more on financing the project by the Project owner and the General Contractor at the start up, with this, the General Contractor or the Seller for other work packages will finance the initial set up and will be paid in a monthly progress billing, the owner will provide the assistance in funding in terms of down payment upon award of the contract, then the succeeding payment will be monthly as submitted by the General Contractor, The General Contractor will have an enough financial resources to carry and sustain the project financial needs, they can also obtain financing or short term loan to sustain initial period of the construction project . Some construction projects eventually failed due to this set up, that's why it needs to be carefully and accurately prequalifying the prospective General Contractor or sellers.

In our template PT-P-7, Pre-qualification criteria of item III , the contracting capability is one of the crucial factors in determining the financial aspect of the prospective seller/contractor ,you can use your company current system or consult your financial advisor in dealing with financial resources checking.

Templates: Financial Management, Credit line-Bank, Credit line-Supplier

FEASIBILITY STUDY

Is the pre-assessment, evaluation and financial analysis to check the viability of the proposed project. Some projects ,feasibility study has been used in the early stage like conceptualization stage or the first phase of the project , other companies they treat feasibility study as standalone project and other project owners needs an improvement or a new facility and to be started by feasibility study. On the contractor side, when they finance a long term projects, contractors usually created feasibility study to check if the project is profitable.

Common Construction Project Phases:

    Pre-construction phase
    Construction phase
    Post-construction phase

Other companies or depending on the complexity of the project they consider this project phase:

    Concept phase (commonly they created feasibility study)
    Planning and Development phase
    Design phase
    Construction phase
    Turn over phase

Template: Feasibility Study

Hot Tip

- Traditional construction like in our assumption here, owner/sponsor will provide a down payment upon award of the contract, down payment in the form of percentage of the total project amount must be clearly specified in the contract.
- The release of down payment must clearly specified in the contract.
- Whenever in the delay of the release of the down payment, what is the capability of the general contractor to finance?. Also , when monthly progress billing is delay, the contractor is capable to sustain project expenditures. *Please consult your financial advisor in dealing with financial resource.*

Key Point

- Ask the prospective bidder for possible additional fund resources , like bank certificates, bank credit line , financing certificate, supplier credit line certificates (must be checked for what they provided)

- If there are delays on the issuance of down payment and the monthly progress billing it must be communicated properly.

- Our assumption here is on traditional construction, where in the owner will provide a down payment and the monthly progress billing, The contractor will start the construction prior to release of NTP (Notice To Proceed).

CLAIMS MANAGEMENT

Claims Management is another group of processes applied for construction projects, this is how to prevent and handle claims into construction projects, there is an originator of claims and the other is the opposing party of claims and it is something a demand for payables to change order, result to activity, directions and demand for time extension.

Templates: Claim Statement , Claim Hour Computation

Hot Tip
- Possible claim such as time extension due to force majeure (acts of nature - earthquakes, typhoon etc.) must be specified in the construction contract (*get an advice to your legal advisor*)

- Provision on the disputes and arbitration must be in the construction contract, check the country of your project's about the agencies for disputes and arbitration condition.

Key Point
- Construction project must be broken down to specific details and as complete as possible to minimize possible future claims.

- No additional or extra works is to be made without proper approval (Change Order process is important here)

		Company Address :
Your Logo here!		Telephone :
		Fax :
		Email : info@
		Website : www.

	Project Management Plan	Date Prepared : 02/28/15	Template ID: PT-P-001
Project	Name of the Project	**Revision :** 000	**Pages :** 1 of 1
Project Package	Name of the package / phase of the project		
Project Location	Name of the location of the project	**Reference No. :** CM-PEC-D-0004-2015	

Project Management Plan Composition

Plans

Scope management plan
Cost management plan
Schedule management plan
Quality management plan
Communications management plan
Procurement management plan
Risk management plan
Human resource plan
Integrated change control plan
  ➢ Change control plan

Baselines

Scope baseline
Schedule baseline
Cost baseline
Quality baseline

	Change Control Plan	Company Address : Telephone : Fax : Email : info@ Website : www.	
		Date Prepared : 02/28/15	Template ID: PT-P-002
Project	Name of the Project	Revision : 000	Pages : 1 of 1
Project Package	Name of the package / phase of the project		
Project Location	Name of the location of the project	Reference No. : CM-PEC-D-0005-2015	

Change Control Board:

Name	Role	Responsibility	Organization	Authority Level

Changes:

Baseline	Description	Impact
Scope		
Schedule		
Cost		
Quality		

Process: *(Establish the process you used or create a process flow you adopted)*

	Company Address :
Your Logo here!	Telephone :
	Fax :
	Email : info@
	Website : www

Requirements Documentation and Traceability

		Date Prepared : 02/28/15	Template ID: PT-SC-001
Project	Name of the Project	Revision : 000	Pages : 1 of 1
Project Package	Name of the package / phase of the project		
Project Location	Name of the location of the project	Reference No. : CM-PEC-D-0005-2015	

ID	Category	Project Requirements Description	Stakeholder	Priority	Impact / connects to Project Purpose	Source	Test / Verification	Acceptance Criteria	Approval	Status

	Project Scope Statement		Company Address : Telephone : Fax : Email : info@ Website : www.	
			Date Prepared : 02/28/15	Template ID: PT-SC-002
Project	Name of the Project		Revision : 000	Pages : 1 of 1
Project Package	Name of the package / phase of the project			
Project Location	Name of the location of the project		Reference No. : CM-PEC-D-0007-2015	

SCOPE

ID	Category	Description

Deliverables

ID	Category	Description	Acceptance Criteria

Project Milestone:

Project Schedule:

Exclusion:

Assumption:

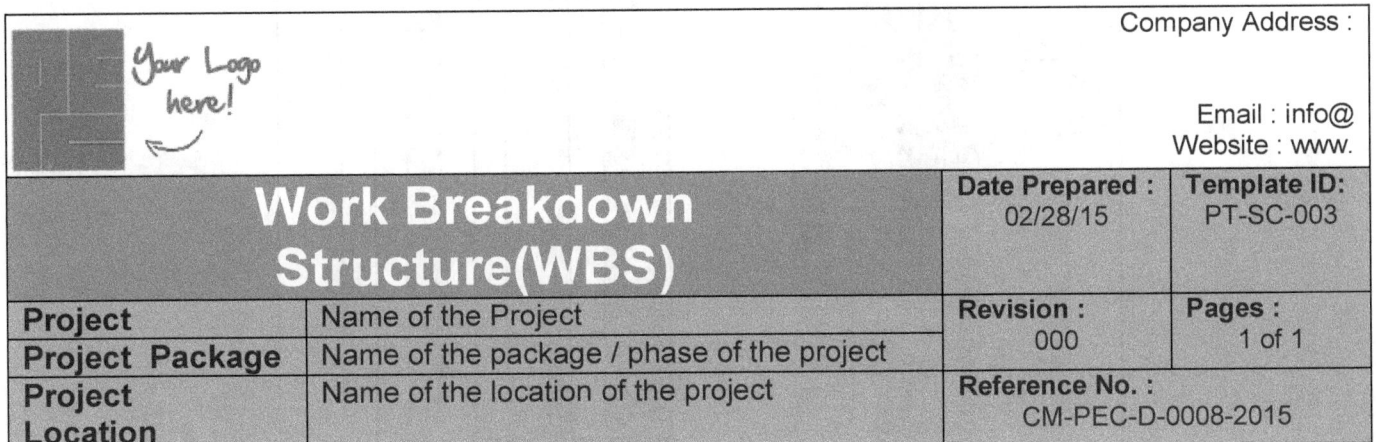

Project	Name of the Project	Revision : 000	Pages : 1 of 1
Project Package	Name of the package / phase of the project		
Project Location	Name of the location of the project	Reference No. : CM-PEC-D-0008-2015	

Work Breakdown Structure(WBS) — Date Prepared: 02/28/15 — Template ID: PT-SC-003

HIERARCHICAL CHART FORMAT

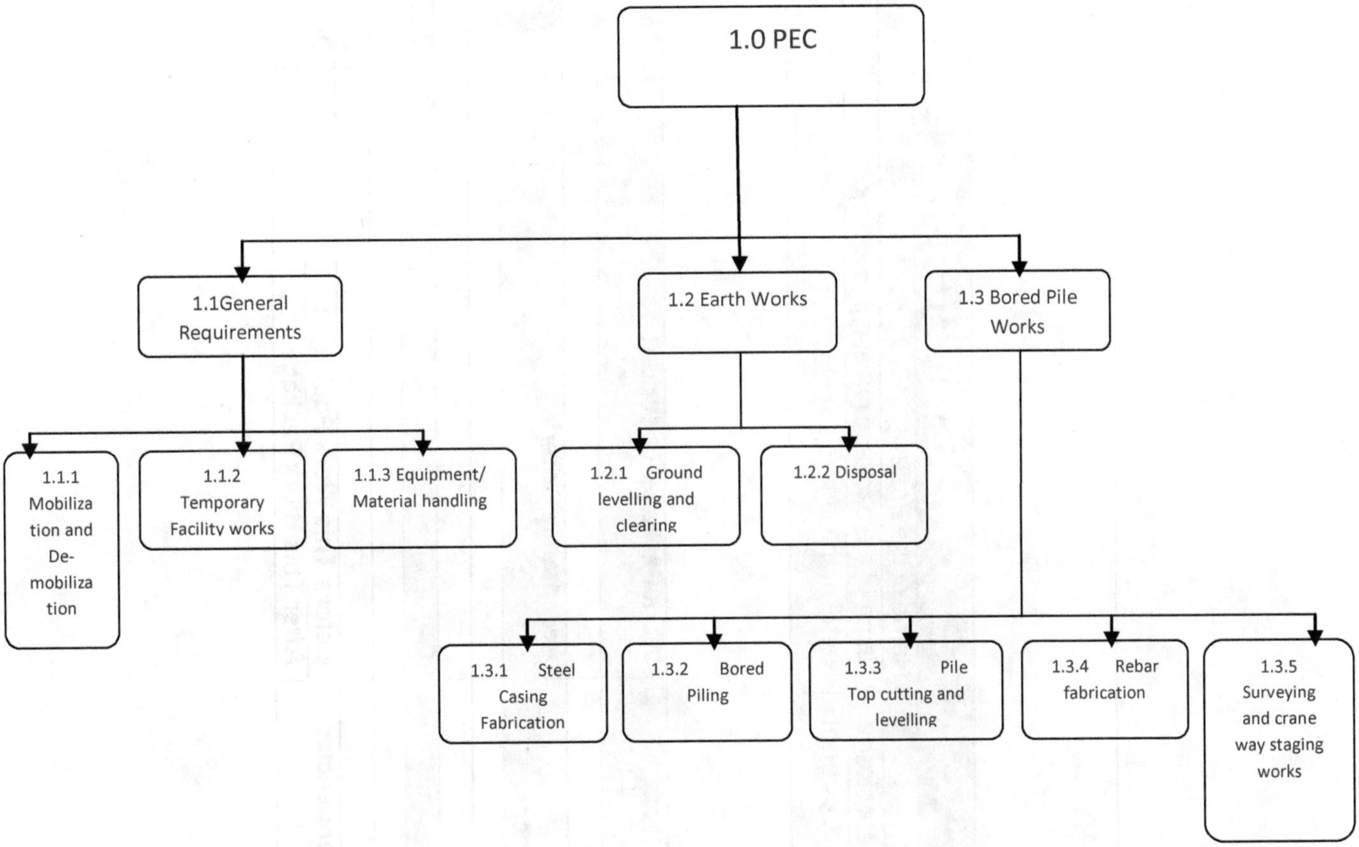

```
1.0 PEC
├── 1.1 General Requirements
│   ├── 1.1.1 Mobilization and De-mobilization
│   ├── 1.1.2 Temporary Facility works
│   └── 1.1.3 Equipment/Material handling
├── 1.2 Earth Works
│   ├── 1.2.1 Ground levelling and clearing
│   └── 1.2.2 Disposal
└── 1.3 Bored Pile Works
    ├── 1.3.1 Steel Casing Fabrication
    ├── 1.3.2 Bored Piling
    ├── 1.3.3 Pile Top cutting and levelling
    ├── 1.3.4 Rebar fabrication
    └── 1.3.5 Surveying and crane way staging works
```

WBS Dictionary

Company Address :	
Telephone :	
Fax :	
Email : info@	
Website : www.	

Date Prepared : 02/28/15	Template ID: PT-SC-004
Revision : 000	Pages : 1 of 2
Reference No. : CM-PEC-D-0009-2015	

Project	Name of the Project
Project Package	Name of the package / phase of the project
Project Location	Name of the location of the project

Work Package ID	Work Package Description	Control Account ID	Responsible (Team /Group / Individual)
Quality Requirements	Acceptance Criteria	Technical References	Status
Duration		Milestones 1. 2.	
Interdependencies	Before This Work Package :	Due Date :	
	After This Work Package :		

WBS Dictionary

Company Address :
Telephone :
Fax :
Email : info@
Website : www.

Date Issue : 02/28/15	Pages : 2 of 2
Reference No. : CM-PEC-D-0009-2015	
Revision No. : 000	

Cost Estimates :

ID	Activity Description	Unit	Quantity	Unit Rate				Total Amount	Resources
				Materials	Labor	Others	Total		

Approved by : Signature:

Project Manager's Name , Date Signed

			Company Address : Telephone : Fax : Email : info@ Website : www.	
Activity List			Date Prepared : 02/28/15	Template ID: PT-T-001
Project	Name of the Project		Revision : 000	Pages : 1 of 1
Project Package	Name of the package / phase of the project			
Project Location	Name of the location of the project		Reference No. : CM-PEC-D-0010-2015	

ID	WBS	Work Package	Activity	Description
1.1.2	General Requirements	Temporary Facility Works	Lay outing and Survey	
			Footing excavation	
			Installation of Pedestal	
			Installation of Steel Containers (as built in office)	
			Ceiling works	
			Installation of insulation for walls and ceiling	

Your Logo here!	**Activity Attributes**	Date Prepared : 02/28/15	Template ID: PT-T-002
		Revision : 000	Pages : 1 of 1
		Reference No.: CM-PEC-D-0011-2015	

Company Address :
Telephone :
Fax :
Email : info@
Website : www.

Project	Name of the Project
Project Package	Name of the package / phase of the project
Project Location	Name of the location of the project

WBS : *(Name and ID of the WBS work)*		Work Package : *(Name and ID of the work)*		Date Prepared :	
ID	Activity Description	Predecessor Activity :	Successor Activity :	Constraints (date)	
		Logical Relationships	Logical Relationships	Assumption :	
		Leads / Lags	Leads / Lags	Where to Make this Activity :	
		Resource Requirements			
		Materials	Equipment	People	
		Quantity / Description	Quantity / Description	Quantity / Description	
		Responsible for This Activity			
		Name :		Location:	

		Company Address :
		Telephone :
		Fax :
		Email : info@
		Website : www.

Milestone List

		Date Prepared: 02/28/15	Template ID: PT-T-003
Project	Name of the Project	**Revision :** 000	**Pages :** 1 of 1
Project Package	Name of the package / phase of the project		
Project Location	Name of the location of the project	**Reference No. :** CM-PEC-D-0012-2015	

ID	Description of Activity	Milestone Started	Milestone Finished
1.1 S	General Requirements - Started	May 01, 2013	
1.1.1	Mobilization		May 14, 2013
1.1.2	Temporary Facilities Installation/erection		June 21, 2013
1.1.3	Material Handling and logistics		June 21, 2013
1.1.4	Demobilization		June 28, 2013
1.1 F	General Requirements - Finished		June 28, 2013
1.2 S	Earth Works- Started	May 03, 2013	
1.2.1	Clearing, grabbing and trimming		May 21, 2013
1.2.2	Hauling and disposal of excavated and cleared materials		May 21, 2013
1.2 F	Earth Works- Finished		June 28, 2013
1.3 S	Load Bearing Elements and Foundation -Started	May 03, 2013	
1.3.1	Steel Casing Fabrication		May 21, 2013
1.3.2	Bored Pile installation and Activities		June 21, 2013
1.3.3	Top of Piles trimming		June 21, 2013
1.3.4	Rebar Case fabrication		June 14, 2013
1.3.5	Survey and Lay-out		June 14, 2013
1.3 F	Load Bearing Elements and Foundation -Finished		June 21, 2013

This Milestone lists is for illustrative only and not intended to portray any construction projects, scope and description or not representing any specific ways to organize, plan of any type of construction projects.

Milestone Schedule

Company Address :	
Telephone :	
Fax :	
Email : info@	
Website : www.	

Date Prepared :	Template ID:
02/28/15	PT-T-004
Revision :	**Pages :**
000	1 of 1
Reference No. :	
CM-PEC-D-0013-2015	

Project	P.E.C School Building
Project Package	Bored Piling Works
Project Location	Name of the location of the project

WBS ID	Description of Activity	May			June		
1.1	General Requirements - Started	◆ 05/01					
1.1.1	Mobilization - Completed		◆ 05/14				
1.1.2	Temporary Facilities Installation/erection - Completed			◆ 05/21			
1.1.3	Material Handling and logistics - Completed			◆ 05/21			
1.2.1	Clearing, grabbing and levelling - Completed					◆ 06/21	
1.2.2	Hauling and disposal - Completed					◆ 06/21	
1.3	Load Bearing Elements and Foundation - Finished						◆ 06/28
1.4	Finished						

This Milestone Schedule is for illustrative only and not intended to portray any construction projects , scope , description or not representing any specific ways to organize , plan of any type of construction projects.

Network Diagram

Project	Name of the Project
Project Package	Name of the package / phase of the project
Project Location	Name of the location of the project

Date Prepared :	02/28/15
Revision :	000
Reference No. :	CM-PEC-D-0014-2015

Template ID:	PT-T-005
Pages :	1 of 1

Company Address :
Telephone :
Fax :
Email : info@
Website : www.

ID	Activity Description
1.0	Start
2.0	Mobilization
3.0	Temporary Facilities Installations
4.0	Clearing
5.0	Survey and Lay out
6.0	Fabrication of Steel casing
7.0	Hauling/Disposal of Cleared/excavated Material
8.0	Fabrication of Rebars
9.0	Bored Piling Works
10.0	Demobilization
11.0	End

This illustration , List is for illustrative only and not intended to portray any construction projects , scope , description or not representing any specific ways to organize , plan of any type of construction projects.

www.ConstructionProjectManagementPro.com

		Company Address :
		Telephone :
		Fax :
		Email : info@
		Website : www.

	Analogous Estimating - Time	**Date Prepared:** 02/28/15	**Template ID:** PT-T-006
Project	Name of the Project	**Revision :** 000	**Pages :** 1 of 1
Project Package	Name of the package / phase of the project		
Project Location	Name of the location of the project	**Reference No. :** CM-PEC-D-0015-2015	

WBS ID	Activity Description	Previous		Present	
		Material Quantity (A)	Duration (B)	Material Quantity (C)	Duration Estimates (B) x (C) / (A)
	Concreting of Floor Slab	10cum.			

		Company Address :	
		Telephone :	
		Fax :	
		Email : info@	
		Website : www.	

Parametric Estimating - Time

		Date Prepared : 02/28/15	Template ID: PT-T-007
Project	Name of the Project	Revision : 000	Pages : 1 of 1
Project Package	Name of the package / phase of the project		
Project Location	Name of the location of the project	Reference No. : CM-PEC-D-0016-2015	

WBS ID	Activity Description	Material Quantity (A)		Productivity rate (Unit per Time) (B)	Duration Estimates (Time) (A) / (B)
		Quantity	Unit		
	Tiles installation	10	sq m	9 sqm/hr	1.11 hours

Project Schedule

Company Address :	
Telephone :	
Fax :	
Email : info@	
Website : www.	

Date Prepared : 02/28/15	Template ID: PT-T-008
Revision : 000	Pages : 1 of 1
Reference No.: CM-PEC-D-0017-2015	

Project	Name of the Project
Project Package	Name of the package / phase of the project
Project Location	Name of the location of the project

WBS ID	Task Name	Duration	Start	Finish
1	Bored Piling Project	56 days?	Thu 4/25/13	Thu 7/11/13
1.1	General Requirements	47 days	Thu 4/25/13	Fri 6/28/13
1.1.1	Mobilization	13 days	Thu 4/25/13	Mon 5/13/13
1.1.2	Temporary Facilities Installation/erection	21 days	Thu 4/25/13	Thu 5/23/13
1.1.3	Material Handling and logistics	42 days	Thu 5/2/13	Fri 6/28/13
1.1.4	Demobilization	14 days	Mon 6/10/13	Thu 6/27/13
1.2	Earth Works	47 days	Thu 4/25/13	Fri 6/28/13
1.2.1	Trimming and clearing	43 days	Thu 4/25/13	Fri 6/28/13
1.2.2	Hauling of excavated and cleared materials	14 days	Tue 4/30/13	Fri 5/17/13
1.3	Load Bearing and Foundation Works	56 days	Thu 4/25/13	Thu 7/11/13
1.3.1	Steel casing fabrication	21 days	Thu 4/25/13	Thu 5/23/13
1.3.2	Bored Piling	35 days	Thu 5/2/13	Wed 6/19/13
1.3.3	Top of Pile cutting off	21 days	Wed 5/22/13	Wed 6/19/13
1.3.4	Fabrication of Rebar Case	35 days	Fri 5/24/13	Thu 7/11/13
1.3.5	Survey and Lay outing	35 days	Fri 5/24/13	Thu 7/11/13
1.4	Sign off	1 day?	Thu 7/11/13	Thu 7/11/13

This illustration is for illustrative only and not intended to portray any construction projects, scope, description of any type of construction projects.

		Company Address :
		Telephone :
		Fax :
		Email : info@
		Website : www.

Analogous Estimating - Cost		Date Prepared : 02/28/15	Template ID: PT-C-001
Project	Name of the Project	Revision : 000	Pages : 1 of 1
Project Package	Name of the package / phase of the project		
Project Location	Name of the location of the project	Reference No. : CM-PEC-D-0018-2015	

WBS ID	Activity Description	Unit	Quantity (A)	Unit Rate(Amount / Unit) (example; $ /cum)				Amount ($ or any unit currencies)
				Material (B)	Labor (C)	Others (contingency) (D)	Total (E) B+C+D	(A) x (E)
	Concreting of Floor Slab	cu.m.	10					

The tabulation above is based on the assumption of the following;

- Unit Rate
    - Derived from previous similar projects
    - Is the value of Amount ($, peso, euro)per unit (cubic meter, linear meter, square foot etc.)
    - Expressed in$ / linear meter, euro /square foot etc.

				Company Address :
				Telephone :
	Your Logo here!			Fax :
				Email : info@
				Website : www.

Parametric Estimating - Cost		Date Prepared : 02/28/15	Template ID: PT-C-002
Project	Name of the Project	Revision : 000	Pages : 1 of 1
Project Package	Name of the package / phase of the project		
Project Location	Name of the location of the project	Reference No. : CM-PEC-D-0019-2015	

WBS ID	Activity Description	Unit	Quantity (A)	Unit Rate(Amount / Unit) (example; $ /cu.m)				Amount ($ or any unit currencies)
				Material	Labor	Others (contingency)	Total	
				(B)	(C)	(D)	(E) B+C+D	(A) x (E)
	Concreting of Floor Slab	cu.m.	10					

The tabulation above is based on the assumptions of the following;

- Unit Rate For Materials (example; $ / cubic meters)
    - Current price per unit given by vendors or updated Amount per unit from commercial data.
    - Derived from detailed estimate using direct counting of the materials, applying formulas considering a portion of area or as a whole.
    - Is the value of Amount ($, peso, euro)per unit (cubic meter, linear meter, square foot etc.)
    - Expressed in $ / linear meter, euro /square foot etc.
- Unit Rate For labour (example; $ / cubic meters, Salary per unit accomplished)
    - Current labor amount per unit given by vendors or updated Amount per unit from commercial data or industry pricing.
    - Derived from detailed estimate using labor rates, applying formulas, established parameters like man-hours (MH).

Cost Performance Baseline - (S-Curve)

Company Address :	
Telephone :	
Fax :	
Email : info@	
Website : www.	

Date Prepared : 02/28/15	Template ID: PT-C-003
Revision : 000	Pages : 1 of 1
Reference No. : CM-PEC-D-0019-2015	

Project	Name of the Project
Project Package	Name of the package / phase of the project
Project Location	Name of the location of the project

WBS ID	Task Name	Duration	Start	Finish
1	Bored Piling Project	56 days?	Thu 4/25/13	Thu 7/11/13
1.1	General Requirements	47 days	Thu 4/25/13	Fri 6/28/13
1.1.1	Mobilization	13 days	Thu 4/25/13	Mon 5/13/13
1.1.2	Temporary Facilities Installation/erection	21 days	Thu 4/25/13	Thu 5/23/13
1.1.3	Material Handling and logistics	42 days	Thu 5/2/13	Fri 6/28/13
1.1.4	Demobilization	14 days	Mon 6/10/13	Thu 6/27/13
1.2	Earth Works	47 days	Thu 4/25/13	Fri 6/28/13
1.2.1	Trimming and clearing	43 days	Thu 4/25/13	Fri 6/28/13
1.2.2	Hauling of excavated and cleared materials	14 days	Tue 4/30/13	Fri 5/17/13
1.3	Load Bearing and Foundation Works	56 days	Thu 4/25/13	Thu 7/11/13
1.3.1	Steel casing fabrication	21 days	Thu 4/25/13	Thu 5/23/13
1.3.2	Bored Piling	35 days	Thu 5/2/13	Wed 6/19/13
1.3.3	Top of Pile cutting off	21 days	Wed 5/22/13	Wed 6/19/13
1.3.4	Fabrication of Rebar Cage	35 days	Fri 5/24/13	Thu 7/11/13
1.3.5	Survey and Lay outing	35 days	Fri 5/24/13	Thu 7/11/13
1.4	Sign off	1 day?	Thu 7/11/13	Thu 7/11/13

You can plot the COST (accumulated cost of all the items on a given time) vs. TIME

This illustration is for illustrative only and not intended to portray any construction projects, scope and description or not representing any specific ways to organize, plan of any type of construction projects.

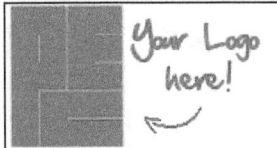

	Quality Management Plan		Date Prepared: 02/28/15	Template ID: PT-Q-001
Project	Name of the Project		Revision : 000	Pages : 1 of 1
Project Package	Name of the package / phase of the project			
Project Location	Name of the location of the project		Reference No. : CM-PEC-D-0021-2015	

Stakeholders:

Name	Organization	Role	Responsibilities	Duty Date
Pier John	Pier Engineering and Consultants			

Quality Assurance Plan:

Quality Control Plan:

Quality Improvement Plan:

Deliverables:

Category	Deliverables	Requirements	Reference	Comments (Date when to measure)

Quality Metrics:

Quality Reports and documents:

Quality Metrics

Company Address :	
Telephone :	
Fax :	
Email : info@	
Website : www.	

		Template ID: PT-Q-002
Date Prepared : 02/28/15		
Revision : 000		Pages : 1 of 3
Reference No. : CM-PEC-D-0022-2015		

Project	Name of the Project
Project Package	Name of the package / phase of the project
Project Location	Name of the location of the project

ID	Category	Item	Description	Method of Measurements	Metrics	Reference	Statistical Sampling
PP-01	Bored Pile	Concrete	Cast in place concrete pile (bored pile)	Compressive Test	3,000 psi (@28 days)	Structural Plan	• 5 Sets of Cylinder at 3 piles / day
				Slump	100mm (max.)	Structural Plan / Structural Specification	• For each batch of concrete • 6 cu.m which ever is lesser
		Reinforcing Steel	Bored pile reinforcing steel	Tensile (Fy) 12mm dia. Bar and larger	413.7 Mpa (60 ksi)		2.5 Tons per diameter per kind
				10mm dia bar And smaller	276 Mpa (40 ksi)		2.5 Tons per diameter per kind
				Bending	No Crack		2.5 Tons per diameter per kind
PP-02e	Concrete Works, supply, fabrication, delivery and erection	Reinforcing Steel	Reinforcing Steel	Tensile (Fy) 12mm dia. Bar and larger	413.7 Mpa (60 ksi)	Structural Plan / Structural Specification	2.5 Tons per diameter per kind
				10mm dia bar And smaller	276 Mpa (40 ksi)		2.5 Tons per diameter per kind
				Bending	No Crack		2.5 Tons per diameter per kind

Quality Metrics

Company Address :
Telephone :
Fax :
Email : info@
Website : www.

Date Issue :	02/28/15
Pages :	2 of 3
Reference No. :	CM-PEC-D-0022-2015
Revision No. :	000

ID	Category	Item	Description	Method of Measurements	Metrics	Reference	Statistical Sampling
PP-02e	Concrete Works, supply, fabrication, delivery and erection	Concrete	Curbs and slab on grade	Compressive Test	3,000 psi (@ 28 days)	Structural Plan / Structural Specification	• 3 Sets of 5 Samples per day of pouring or 150 cu.m. concrete poured Or 500 sq.m slab or walls • For each batch of concrete (min) 10 cu. Yard (max) • For each batch of concrete (min) 10 cu. Yard (max)
				Slump	100 mm (max,)		Same as above
				Temperature	35 degrees Celsius (max)		Same as above
			Pile Cap/ Tie Beam/ Footing	Compressive Test	3,000 psi (@ 28 days)		Same as above
				Slump	100 mm (max,)		Same as above
				Temperature	35 degrees Celsius (max)		Same as above
			Beam /Slab	Compressive Test	3,000 psi (@ 28 days)		Same as above
				Slump	100 mm (max,)		Same as above
				Temperature	35 degrees Celsius (max)		Same as above

Quality Metrics

Company Address:
Telephone:
Fax:
Email: info@
Website: www.

Pages: 3 of 3
Date Issue: 02/28/15
Reference No.: CM-PEC-D-0022-2015
Revision No.: 000

ID	Category	Item	Description	Method of Measurements	Metrics	Reference	Statistical Sampling
PP-02e	Concrete Works, supply, fabrication, delivery and erection	Concrete	Column Shear wall	Compressive Test	3,000 psi (@ 28 days)		Same as above
				Slump	100 mm (max.)		Same as above
				Temperature	35 degrees Celsius (max)		Same as above
			Lean Concrete	Compressive Test	1,000 psi (@ 28 days)		Same as above
				Slump	100 mm (max.)		Same as above
				Temperature	35 degrees Celsius (max)		Same as above

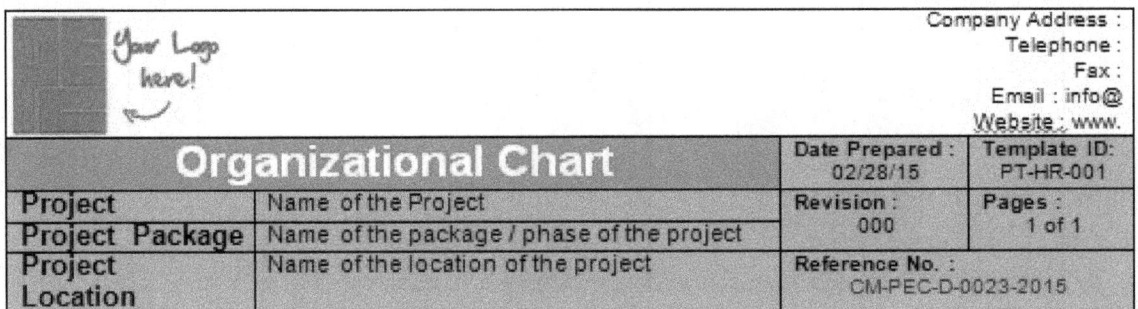

ORGANIZATIONAL CHART - Project Management Team

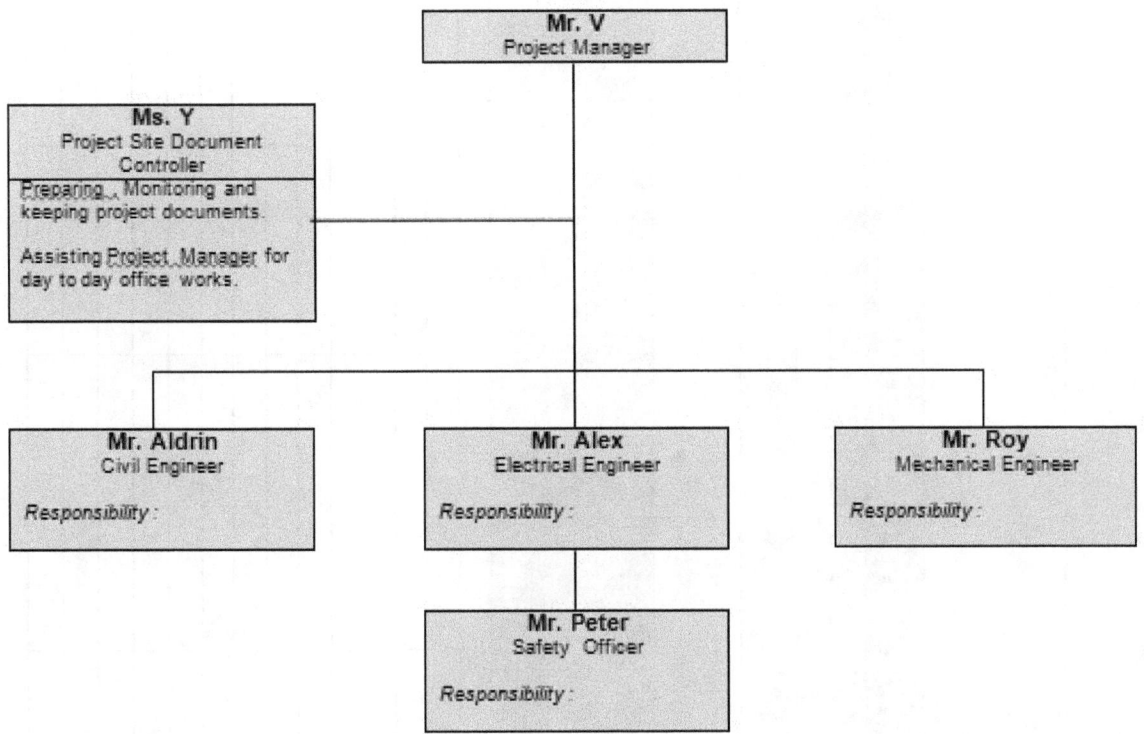

Responsibility Assign Matrix

Company Address :		
Telephone :		
Fax :		
Email : info@		
Website : www		

Date Prepared :	02/28/15	Template ID: PT-HR-002
Revision :	000	Pages : 1 of 1
Reference No. :		CM-PEC-D-0024-2015

Project	Name of the Project
Project Package	Name of the package / phase of the project
Project Location	Name of the location of the project

R = Responsible, Person that performs the work
A = Accountable, Person that capable to answer that the requirements of the activity or work packages is acceptable
C = Consult, Person that has necessary data or information to execute and finish the required activity or work package
I = Inform, Person's to inform when the activity or work packages is complete

ID	Category	Activity	Description	Name 1	Name 2	Name 3	Name 4

www.ConstructionProjectManagementPro.com

Your Logo here!	**Human Resource Plan**		**Date Prepared:** 02/28/15	**Template ID:** PT-HR-003
Project	Name of the Project		**Revision :** 000	**Pages :** 1 of 4
Project Package	Name of the package / phase of the project			
Project Location	Name of the location of the project		**Reference No. :** CM-PEC-D-0025-2015	

Company Address :
Telephone :
Fax :
Email : info@
Website : www.

Roles and Responsibility:

Roles	Responsibility	Authority	Competence / Ability	Duty Schedule	
				Start	End

Project Organizational Chart: (*see separate sheet attached – Organizational Chart*)

		Company Address :
Your Logo here!		Telephone : Fax : Email : info@ Website : www.
Human Resource Plan		Date Issue : 02/28/15	Pages : 2 of 4
		Reference No.: CM-PEC-D-0025-2015
		Revision No. : 000

Staffing Management Plan:

Staff Acquisition						
Category	Role	Status		Location	Cost	
		In house	outsource		weekly	Equivalent monthly

Human Resource Histogram

Human Resource Plan

Date Issue : 02/28/15	Pages : 3 of 4
Reference No.: CM-PEC-D-0025-2015	
Revision No. : 000	

Company Address :
Telephone :
Fax :
Email : info@
Website : www.

Staff Release	

Training		Awards	
Date		Date	

Human Resource Plan

Date Issue : 02/28/15	**Pages :** 4 of 4
Reference No.: CM-PEC-D-0025-2015	
Revision No. : 000	

Compliance
(Regulations, Contract and Policy)

Human Resource Safety Provisions

Your Logo Here!

Company Address:
Telephone:
Fax:
Email: info@
Website: www.

Template ID : PT-HR 004
Revision: 000

Reference No. : CM-PEC-D-0026-2015   Date: February 28, 2015

To : **NAME OF HIRED EMPLOYEE**

Project : **DESIGNATED PROJECT**

Subject : **EMPLOYMENT CONTRACT - TEMPORARY AS PROJECT TO PROJECT BASIS ONLY**

This serves as your temporary employment contract, bounded by the following terms and condition.

I. DESIGNATION : Specify role and of the hired employee and which project will he be working on.

II. DURATION : Specify duration of the project, including the provision for termination.

III. COMPENSATION : Specify amount of salary per week or per month that he will be getting, including provision of benefits and allowances.

IV. RESPONSIBILITIES : Specify list of employee's responsibilities as required to the project.

Signed and agreed upon this 1st day of September, 2013, at New Hampshire Building, New York.

Company Name

By: Conforme:

NAME OF COMPANY REPRESENTATIVE   **NAME OF EMPLOYEE**
President   Employee

This Temporary Employment Contract is for illustrative purposes only and not intended to portray any actual construction contracts for employment. To deal with this, please consult your Legal Department or Human Resource Department.

Your Logo Here!

Company Address:
Tel :0000000
Fax : 0000000
Email : info@
Website: www

Template ID: PT-HR-005
Revision: 000

Reference No. : CM-PEC-D-0027-2015

Date: February 28, 2015

To	:	**PROJECT MANAGER**
Project	:	**NAME OF THE PROJECT**
Subject	:	**CONTRACT EXPIRATION NOTICE**

The contracts of staff members listed below are due to expire.

Staff Member's Name	Role	Project Assignment	Date of Effectivity

Please advise us on or before thirty (30) days prior to their contracts status.

Prepared by:

Signature:

_____
Administrative Officer

Your Logo Here!

Company Address:
Tel :0000000
Fax : 0000000
Email : info@
Website: www

Template ID: PT-HR-006
Revision : 000

Reference No. : CM-PEC-D-0028-2015

Date : February 28, 2015

To	: **PROJECT MANAGER**
Project	: **NAME OF THE PROJECT**
Subject	: **CONTRACT EXTENSION REQUEST**

I recommend here with for your approval, the extension of the contract of the following staff members listed below.

Staff Member's Name	Role	Contract Expiration	Proposed Extension		Reason of Extension
			Start	End	

Prepared by: Approved by:

_____
Construction Manager   Project Manager

Cc: Administrative Officer
Project File

Your Logo Here!

Company Address:
Tel :0000000
Fax : 0000000
Email : info@
Website: www

Template ID: PT-HR-007
Revision : 000

Reference No. : CM-PEC-D-0029-2015

Date: February 28, 2015

Attention	:	**STAFF MEMBER'S NAME** Role
Project	:	**Name of the project**
Project Package	:	PP 01 – General Civil and Architectural Works
Location	:	Project Location
Subject	:	**NON RENEWAL OF CONTRACT**

Dear **STAFF MEMBER'S NAME,**

    We would like to inform you that we are not going to renew your contract for the project mentioned above. Your contract will be terminated on ***Date and Time.***

Respectfully Yours,

_____
Signature

_____
Printed Name / Date

Administrative Officer

Noted by:

_____
Signature

_____
Printed Name / Date

Construction / Project Manager

www.ConstructionProjectManagementPro.com

Your Logo Here!

Company Address:
Tel :0000000
Fax : 0000000
Email : info@
Website: www

Template ID: PT-HR-008
Revision: 000

Reference No. : CM-PEC-D-0030-2015 Date: February 28, 2015

Attention	:	**STAFF MEMBER'S NAME**
		Role
Project	:	**Name of the project**
Project Package	:	PP 01 – General Civil and Architectural Works
Location	:	Project Location
Subject	:	**CONTRACT EXTENSION**

Dear **STAFF MEMBER'S NAME,**

Please be advised that we are extending your contract for the project mentioned above. Your contract will be effective **Date and Time.** Previous Contract conditions remain unchanged and your current role will be the same.

Kindly affix your signature with the space provided below for your conformity.

Respectfully Yours,

_____
Printed Name / Date
_____
Construction / Project Manager

Conform;

_____
Printed Name / Date
_____
Staff Member's Role

	Company Address:
	Telephone:
	Fax:
	Email: info@
	Website: www.

Communication Management Plan	**Date Prepared:** 02/28/15	**Template ID:** PT-COM-001
Project — Name of the Project	**Revision:** 000	**Pages:** 1 of 3
Project Package — Name of the package / phase of the project		
Project Location — Name of the location of the project	**Reference No.:** CM-PEC-D-0031-2015	

Information / Message : (Details of information or message to be communicated)

Information Description	Method	Stakeholder					Frequency
		Receiver			Sender		
		Name	Organization	Reason for distribution	Name	Organization	

Resources:

Category	Description	Schedule	Budget	Reference / Provision

Flow charts and Work flows:

Communication Management Plan

Date Issue: 02/28/15	Pages: 3 of 3
Reference No.: CM-PEC-D-0031-2015	
Revision No.: 000	

Glossary:

Terminology	Definition

			Company Address : Telephone : Fax : Email : info@ Website : www.
Risk Management Plan		**Date Prepared :** 02/28/15	**Template ID:** PT-R-001
Project	Name of the Project	**Revision :** 000	**Pages :** 1 of 4
Project Package	Name of the package / phase of the project		
Project Location	Name of the location of the project	**Reference No. :** CM-PEC-D-0032-2015	

Risks Analysis:

Qualitative Risk Analysis

Risks: (use this also as a tool and technique to identify risk, explore more possible risks)

ID	Category			Description
	Internal			
		Funds		
		Costs		
		Time / Schedule		
		Resources		
			Material	
			Equipment	
			Manpower	
		Contractor		
		Sponsor		
	External			
		Suppliers		
		Sub - Contractor		
		Government		
		Regulatory		
		Market		
		Cultural		
		Customer		
		Nature		

	Company Address :
	Telephone :
	Fax :
	Email : info@
	Website : www.

Risk Management Plan

Date Issue : 02/28/15
Pages : 2 of 4
Reference No.: CM-PEC-D-0032-2015
Revision No. : 000

ID	Category			Description
			Weather	
			Acts of Nature	
	Technical			
		Compatibility / Interface		
		Requirements (Technology, Performance)		
		Quality		
		Scope		
	Project Management			
		Knowledge		
		Experience		
		Planning		
			Estimates	
			Communication	
			Human Resource	

Stakeholder:

Roles	Organization	Responsibility

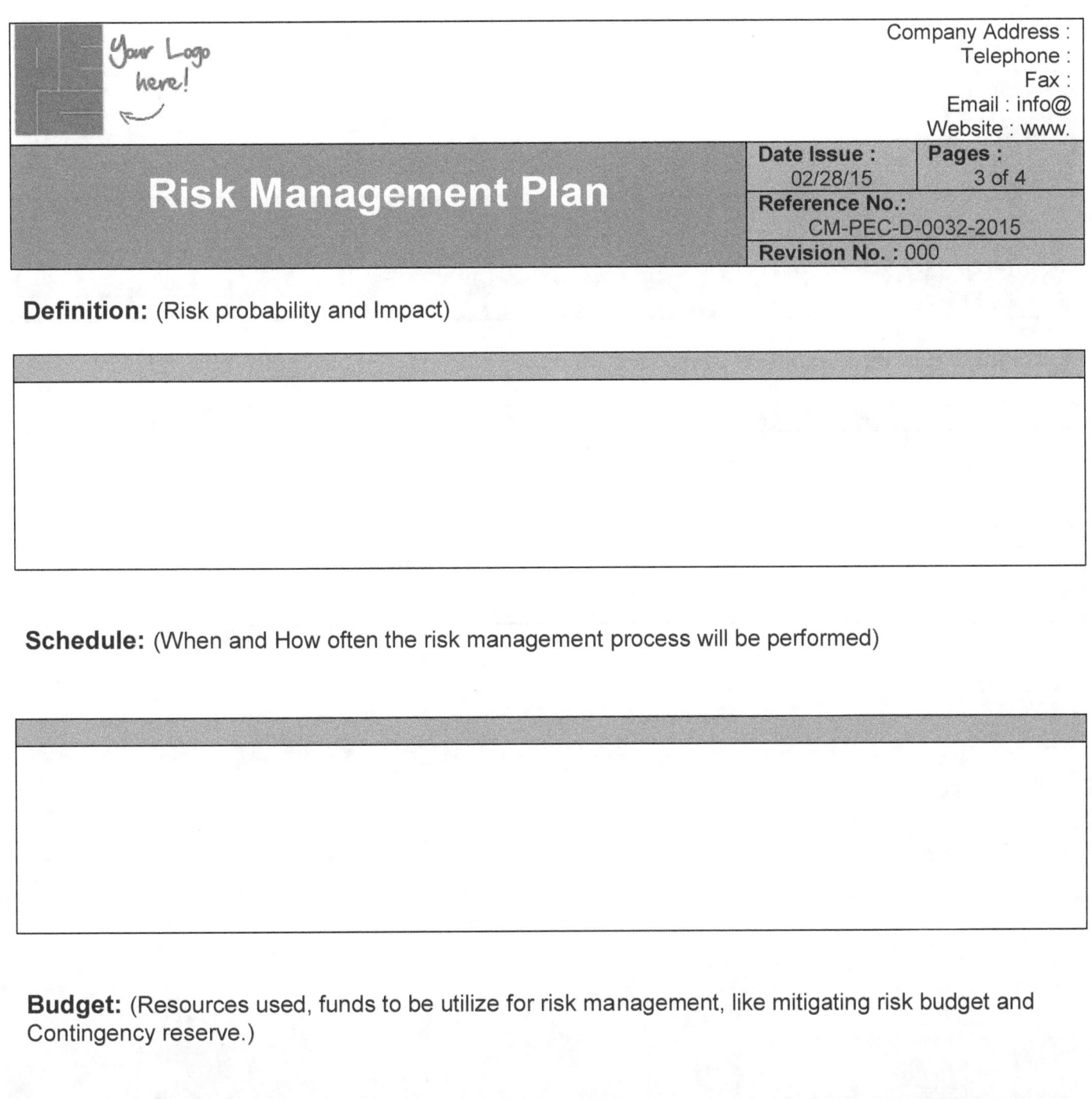

	Risk Management Plan	Date Issue : 02/28/15	Pages : 3 of 4
		Reference No.: CM-PEC-D-0032-2015	
		Revision No. : 000	

Definition: (Risk probability and Impact)

Schedule: (When and How often the risk management process will be performed)

Budget: (Resources used, funds to be utilize for risk management, like mitigating risk budget and Contingency reserve.)

	Company Address :
Risk Management Plan	Telephone :
	Fax :
	Email : info@
	Website : www.

Date Issue : 02/28/15	Pages : 4 of 4
Reference No.: CM-PEC-D-0032-2015	
Revision No. : 000	

Tracking: (How the risks will be audited and recorded)

Risk Register

Company Address :	
Telephone :	
Fax :	
Email - info@	
Website - www.	

Date Prepared :	02/28/15
Revision :	000
Reference No. :	CM-PEC-D-0033-2015
Template ID:	PT-R-002
Pages :	1 of 3

Project	Name of the Project
Project Package	Name of the package / phase of the project
Project Location	Name of the location of the project

ID	Category	Probability Score	Impact Score	Priority Score	Description	Response Owner	Response Date	Status
	Internal							
	Funds							
	Costs							
	Time / Schedule							
	Resources							
	• Material							
	• Equipment							
	• Manpower							
	Contractor							
	Sponsor							

Risk Register

Company Address:
Telephone:
Fax:
Email: info@
Website: www.

Date Issue: 02/28/15	Pages: 2 of 3
Reference No.:	CM-PEC-D-0033-2015
Revision No.: 000	

ID	Category	Probability Score	Impact Score	Priority Score	Description	Response Owner	Response Date	Status
	External							
	Suppliers							
	Sub - Contractor							
	Government							
	Regulatory							
	Market							
	Cultural							
	Customer							
	Nature							
	• Weather							
	• Acts of Nature							

ID	Category	Probability Score	Impact Score	Priority Score	Description	Response Owner	Response Date	Status
	Technical							
	Compatibility/Interface							
	Requirements (technology , performance)							
	Quality							
	Scope							

Risk Register

Company Address:
Telephone:
Fax:
Email: info@
Website: www.

Date Issue: 02/28/15	Pages: 3 of 3
Reference No.: CM-PEC-D-0033-2015	
Revision No.: 000	

ID	Category	Probability Score	Impact Score	Priority Score	Description	Response			Status
						Owner	Date		
	Project Management								
	Knowledge								
	Experience								
	Planning								
	• Estimates								
	• Communication								
	• Human Resource								

Company Address :
Telephone :
Fax :
Email : info@
Website : www.

Owner Supplied Materials

		Date Prepared : 02/28/15	Template ID: PT-P-001
Project	Name of the Project	**Revision :** 000	**Pages :** 1 of 1
Project Package	Name of the package / phase of the project		
Project Location	Name of the location of the project	**Reference No. :** CM-PEC-D-0034-2015	

No.	Description	Brand/Specification	Quantity	Unit	Date Needed on Site	Remarks

	Procurement Management Plan	**Date Prepared:** 02/28/15	**Template ID:** PT-P-002
Project	Name of the Project	**Revision:** 000	**Pages:** 1 of 2
Project Package	Name of the package / phase of the project		
Project Location	Name of the location of the project	**Reference No.:** CM-PEC-D-0034-2015	

Company Address :
Telephone :
Fax :
Email : info@
Website : www.

STAKEHOLDER: (person involved in procurement, manage, documents and executing procurement contracts)

Role	**Responsibility** (Describe the responsibility in the procurement process)	**Authority level** (define the authority level for, costs, technical, decisions, changes, transactions)
Project Manager		
Team		
Procurement Manager		

TYPE OF CONTRACT : (State the contract type to be used)

DOCUMENTATION : (List all the necessary procurement documents and integration process to organization procurement documents, provision for OSME)

ESTIMATES : (independent estimates to be use, evaluation or selection.)

SCHEDULE : (Establish how the seller will integrate their supply or services schedule with the project schedule)

PERFORMANCE : (Establish how the seller will integrate their performance reports and all necessary submittals to procurement process)

WBS : (Establish how the seller will adopt and maintain and develop the project WBS with their integration of supplies or services

BONDS OR INSURANCES: (describe the bonds / insurance to be carried out on the contract as part of mitigating the risk).

ASSUMPTION : (Establish necessary assumption to be used for procurement process.)

CONSTRAINTS : (Establish necessary constraints to be used for procurement process.)

RISKS : (Establish Risks associated with procurement process)

			Company Address :
Your Logo here!			Telephone :
			Fax :
			Email : info@
			Website : www
Procurement Statement of Work		**Date Prepared:** 02/28/15	**Template ID:** PT-P-003
Project	Name of the Project	**Revision :** 000	**Pages :** 1 of 1
Project Package	Name of the package / phase of the project		
Project Location	Name of the location of the project	**Reference No. :** CM-PEC-D-0036-2015	

ID	Category	Item Description	Unit	Quantity

Documents : (Attached detailed documents in a separate sheets, if contained large paper size pages)

Document Number	Description	Number of pages	Reference Number
1	Quantities		
2	Work Schedule		
3	Technical Specification		
4	Reference Drawing		

Request For Information (RFI)		RFI No.	Date Prepared :	Template ID:
		000	02/28/15	PT-P-004
Project	Name of the Project		Revision :	Pages :
Project Package	PP 01 – General Civil and Architectural Works		000	1 of 1
Project Location	Name of the location of the project		Reference No. : CM-PEC-D-0037-2015	

SELLER'S / ARCHITECT'S / ENGINEERS COMPANY NAME
Company Address

ATTENTION   :   SELLER'S / ARCHITECT'S / ENGINEER'SNAME
                                  Role

SUBJECT : *Sample Only :* **Roof Framing Specification**

Description :
Sample Only : May we request the complete description of the chord WT 4x15.5 (in x lb/ft) for the main roof truss. Can we useW8 x 31 (in x lb/ft) and cut into half to get the said section?

Instruction / Comment :

_____
Name and Signature Date

Prepared by:

_____
Signature

_____
Printed Name / Date

_____
Role / Organization or Company

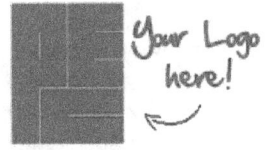

Company Address:
Tel:0000000
Fax: 0000000
Website: www

Template ID: PT-P-005
Revision: 000

Reference No. : CM-PEC-D-0038-2015   Date: February 28, 2015

Seller's / Bidder's Company Name
Address

Attention	:	**Seller's Name or Company representative** Role
Project	:	**Name of the project**
Project Package	:	PP 01 – General Civil and Architectural Works
Location	:	Project Location
Subject	:	INVITATION FOR PREQUALIFICATION OF SELLER

Gentlemen,

We are pleased to invite you to participate in the prequalification of seller/bidder for the **supply of materials, labor, tools, equipment and supervision to complete Project Package 01 (PP 01)** General Civil and Architectural Works for the above mentioned project, below are the documents to be submitted.

Technical;

- Records of previous projects (similar and not similar projects).
- Company background, details and location.
    - Business Permit
    - Contractor's License
    - Company Certifications (ISO etc)
- Personnel to be assign, details and experiences.

Financial;

- Total Current Assets and Liabilities
- Bank Certificates

(Above financial data is to be certified by Accountant or Auditor)

Company Address:
Tel:0000000
Fax: 0000000
Website: www

Reference No. : CM-PEC-D-0038-2015   Date: February 28, 2015

Sealed Technical and Financial documents in two(2) copies each and marked **Application for Prequalification for *Project Name*** must be submitted and stamped received at the office of;

        Attention: **MR. ABC (**Owner's Representative / Document Receiver)
                             Company Address

                          Not later than, 2:00pmon Date _____.

The Project Manager reserves the right to reject any of the application for prequalification or any part thereof, to waive any informality there in and to select qualified seller or bidder that would be found in its opinion to be most reliable to the project manager.

Please signify your interest for pre-qualification for bid by signing on the space below and send back to *Pier Engineering and Consultants* thru email at *info@pierengineeringandconsultants.com,* or fax,*00000000*, not later than 10:00 am, Date _____.

Respectfully Your,

PIER JOHN
Project Manger

Invitation to pre-qualify for bid,
received and confirmed intent by;

_____
Signature

_____
Printed Name / Date

_____
Role / Organization or Company

	Company Address: Tel:0000000 Fax: 0000000 Website: www
	Template ID : PT-P-006 Revision : 000

Reference No. : CM-PEC-D-0039-2015   Date: February 28, 2015

Seller's / Bidder's Company Name
Address

Attention	:	**Seller's Name or Company representative** Role
Project	:	**Name of the project**
Project Package	:	PP 01 – General Civil and Architectural Works
Location	:	Project Location
Subject	:	**INVITATION FOR BID (IFB)**

Gentlemen,

This is to inform you that you have been prequalified and invite you to participate in the bidding for the **supply of materials, labor, tools, equipment and supervision to complete Project Package 01 (PP 01)** General Civil and Architectural Works for the proposed" *name of the project*" located in" project location".

Structural, Architectural plans and all related project documents shall be available for pick up on *"specify time and date"* at the office of;

> **MR. ABC (**Owner's Representative / Designer etc.**)**
> Company Address
> Contact Numbers

Upon payment of non-refundable cash deposit of **USD (Amount in words), $ (Amount in figures).**

Sealed Technical and Financial documents in two(2) copies each and marked " **Proposal For PP-01GeneralCivil and Architectural Works-*Project Name*"** must be submitted and stamped received at the office of;

> Attention: **MR. ABC(**Owner's Representative / Document Receiver**)**
> Company Address
>
> Not later than, 2:00pmon Date _____ .

Company Address:
Tel: 0000000
Fax: 0000000
Website: www

Reference No. : CM-PEC-D-0039-2015   Date: February 28, 2015

The Owner / Sponsor reserves the right to reject any of the Bids or any part thereof, to waive any informality there to award the contract to any of the Bidders / Sellers select that would be found in its opinion, to be most reliable and advantageous to the owner.

A pre bid conference will be scheduled on, "**Date, Day**" at the office of **Company Name of Consultant or Sponsor** at "**Time**". Please prepare your written clarifications prior to this meeting date.

Please signify your interest to bid by signing on the space below and send back to **Company Name of Consultant or Sponsor** thru email at *email address* or fax No.___ not later than 10:00 am, Date _____.

Respectfully Yours,

PIER JOHN
Project Manger

Invitation for bid received and
confirmed intent to bid by ;

_____
Signature

_____
Printed Name / Date

_____
Role / Organization or Company

Company Address :
Telephone :
Fax :
Email : info@
Website : www.

Pre – Qualification Criteria

Project	Name of the Project
Project Package	Name of the package / phase of the project
Project Location	Name of the location of the project

Date Prepared : 02/28/15	Template ID: PT-P-007
Revision : 000	Pages : 1 of 5
Reference No. : CM-PEC-D-0040-2015	

DESCRIPTION	POINTS	BIDDER/SELLER/CONTRACTORS COMPANY X (Name of the Sellers Company)		
		Projects	Project Cost	Equivalent points
I. EXPERIENCE AND CAPABILITY OF FIRM				
a.) Previous Project Records (Similar projects)	30			20
1.) With similar projects in the last 5 years (20 pts.)	20			
2.) Without similar projects but with comparable nature (15 pts.)				
b.) Repeat engagement with the same client/customer as Seller/Contractor	5			5
c.) Geographical consideration for intended project	5			5
Location of office or satellite office is within the region (5 pts.)				
Location of office or satellite office is outside the region (2 pts.)				
Sub - Total I	30			30

Company Address :			
Telephone :			
Fax :			
Email : info@			
Website : www.			
Date Issue : 02/28/15		Pages : 2 of 5	
Reference No. :		CM-PEC-D-0040-2015	
Revision No. : 000			

Pre – Qualification Criteria

II. QUALITY OF PERSONNEL TO BE ASSIGNED	30	NUMBER OF YEARS	30
1. Project Manager / Project Engineer			
1.1 Experience of the Project Manager	20	(example entry)	(example)
15 years and above (20 pts.)		Project Manager - 14 years (put 10 points)	10
10 – 14 years (10 pts.)			
5 – 9 years (5 pts.)			
1.2 Experience of Project Engineer	5	(example entry)	(example)
10 years and above (5 pts.)		Project Engineer - 5 years (put 3 points)	3
5 – 9 years (3 pts.)			
1.2.1 Years in Company	5		
6 years and above (3 pts.)			
5 years and below (2 pts.)			
5 years and below (1 pt.)			

Pre – Qualification Criteria

Company Address :
Telephone :
Fax :
Email : info@
Website : www.

Date Issue : 02/28/15	Pages : 3 of 5
Reference No. : CM-PEC-D-0040-2015	
Revision No. : 000	

Project Engineer		
6 years and above (2 pts.)		
5 years and below (1 pt.)		
Sub - Total II	*30*	*30*

III. CONTRACTING CAPABILITIES	40	Year (latest)	40
1.1 Total Current Assets			
1.2 Total Current Liabilities			
1.3 Net Worth = TCA - TCL			
1.4 Approximate Cost of Project			
1.5 Maximum Contracting Capacity = Net Worth x 10			
1.6 Gross Contracting Capability = MCC x 0.30			
1.7 Equity = Net Worth x 10			
1.8 Net Financial Contracting Capability (NFCC)			

	Company Address :
	Telephone :
Your Logo Here!	Fax :
	Email : info@
	Website : www...

Pre – Qualification Criteria

Date Issue : 02/28/15	Pages : 4 of 5
Reference No. :	CM-PEC-D-0040-2015
Revision No. : 000	

Conditions:

If the Net Financial Contracting Capability (NFCC) is equal to or greater than Estimated Cost of the project, the seller or the bidder is financially qualified.

If the NFCC is less than half of the Estimated Cost of the Project, the seller or the bidder is financially not capable or disqualified.

If the NFCC equals at least half of the Estimated Cost of the Project, the seller's or the bidder's is financially un capability or insufficient. To satisfy the requirements, direct loans, financier or other instruments from banking/financial institutions is needed in an amount equal or at least one and a half (1.5) times the deficiency in net worth. Otherwise, the bidder is financially disqualified.				40

If the NFCC equals at least half of the Estimated Cost of the Project, the seller's or the bidder's is financially un capability or insufficient. To satisfy the requirements, direct loans, financier or other instruments from banking/financial institutions is needed in an amount equal or at least one and a half (1.5) times the deficiency in net worth. Otherwise, the bidder is financially disqualified.	Sub - Total III	40

Company Address :	
Telephone :	
Fax :	
Email : info@	
Website : www.	

Date Issue :	Pages :
02/28/15	5 of 5
Reference No. :	
CM-PEC-D-0040-2015	
Revision No. : 000	

Pre – Qualification Criteria

TOTAL POINTS I+II+III	100			100

To qualify For Bid the Seller /Contractor should have a total points of not less than 75 points.

Prepared by:

Signature :
Name :
Role and Organization :

www.ConstructionProjectManagementPro.com

STEP 3 : EXECUTING

The Executing Process Group is the third process group with the purpose of applying processes to complete the work defined in our project management plan , executing is the coordination of all resources like , people , materials and equipment and the necessary activities in accordance with the specified project management plan.

Flow Chart - Project Management process is not always sequential or performed in identical sequence.

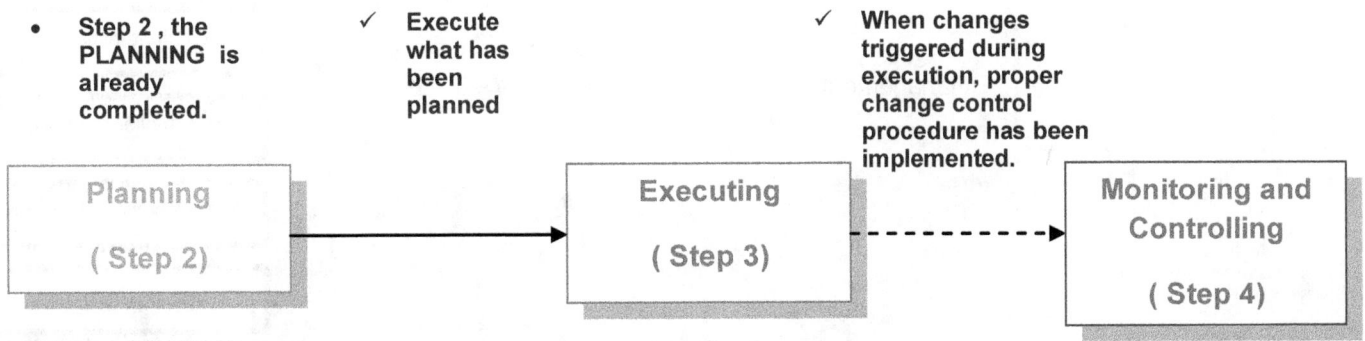

Construction Management: Step-by-Step Templates — Step 3 : Executing

Mapping:

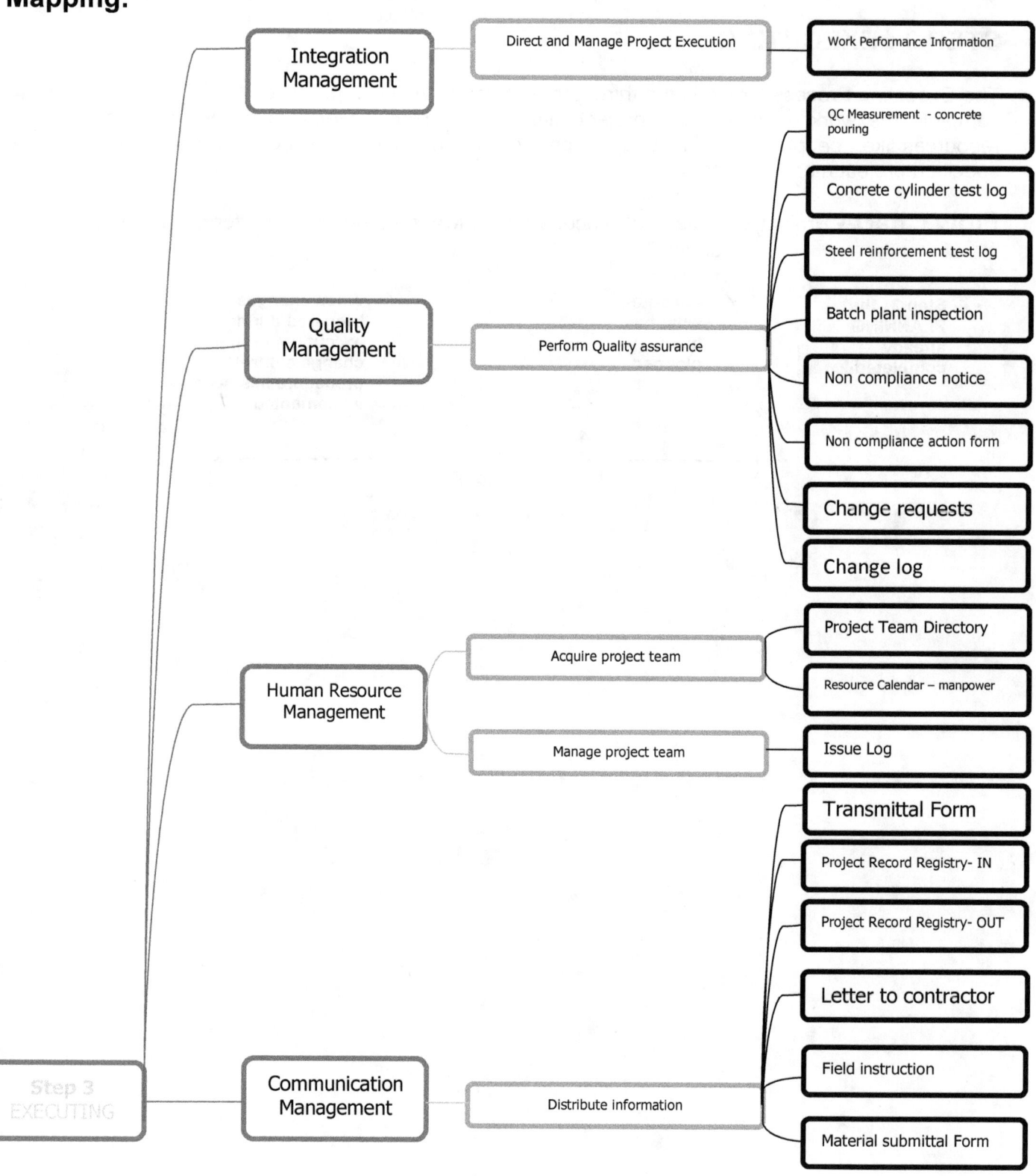

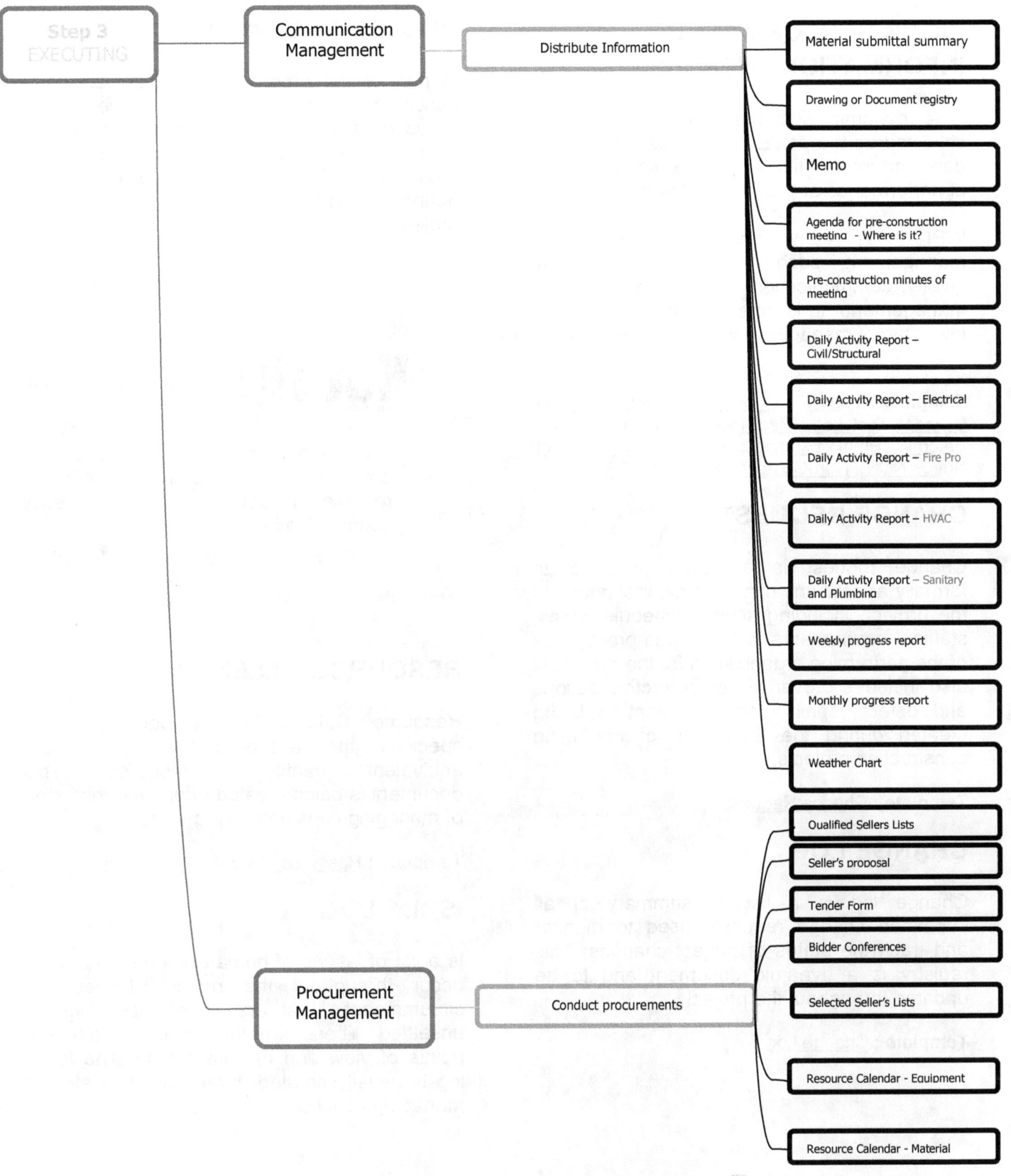

WORK PERFORMANCE INFORMATION (WPI)

Is a dynamic document which is routinely collected as the project progresses. It involves data and information about the actuals results of performance or progress of the project.

WPI has many uses in the project management, such as scope, time, cost, quality, communication, risk, and procurement management. It is often created during the third step of managing construction projects.

Templates : WPI , Concrete Pouring, Concrete Cylinder Test Log, Steel Reinforcement Test Log, Batch Plant Inspection, Non-Compliance Notice(NCN), Non-Compliance Action(NCA)

CHANGE REQUEST

Change request is a document used to formally ask for any minor or major changes in the project, including scope, schedule, costs, staffing requirements, policies, and procedures of the performing organization for the project. It also includes preventive or corrective actions and defect repair. This document is being created during the third step of managing construction projects.

Template : Change Request

CHANGE LOG

Change log is a list or summary of all submitted change requests used to monitor and track the status of project changes. This registry is a dynamic document and to be updated throughout the project.

Template : Change Log

PROJECT TEAM DIRECTORY

Is part of the project staff assignment. This document is a list of members of the project, including the complete information of the member like email, mobile numbers, and organization. This document is being created during the third step of managing construction projects.

- Post site team directory inside the office (contractor, designer, consultants, and owner's representatives) for easy communication.

Template : Project Team Directory,

RESOURCE CALENDAR

Resource Calendar is a document that specifies the availability, schedule, and equivalent quantity of resources. This document is being created during the third step of managing construction projects.

Template : Resource Calendar-Manpower

ISSUE LOG

Is a list of issues of human resource that may occur throughout the project. Issues are circumstances that are under discussion or unsettled. These are the results of different points of view and disagreements. The issue log is usually created during the third step of managing construction projects.

INFORMATION DISTRIBUTION

Information distribution is the process of providing necessary information available to stakeholders, and is often performed throughout the project duration. During the distribution of required information to stakeholders, it is best to track the important information, such as emails and hard copies with transmittal form to be signed by the receiver of the document. The information distribution is being done during the third step of managing construction projects.

Templates : Transmittal Form, Project Records Registry-IN, Project Records Registry-OUT, Letter To Seller, Field Instructions (FI), Material Submittal Form, Material Submittal Summary, Drawing or Document Registry, Memo, Pre-construction Meeting Agenda, Pre-construction Minutes of Meeting

DAILY ACTIVITY REPORT

Daily Activity Report is being used in the construction industry for progress reporting on the daily basis and it is bring up into summary level for project status report , daily activity report is being prepared by team member and to be submitted to project manager as detailed , summary or broad form depending on the needs of the project , this document is very useful for risk , safety , environmental issues and project progress , also , along with this report is the weather chart and being filled up on the daily basis .The daily activity report and project weather chart is being created in the third step or the STEP 3 in managing construction projects

Also included here is the weather chart , for construction project , it is important to record the daily condition of the weather , for future possible construction claims , when time extension is the issue , some will fall under force majeure and one of the references for hours affected will be the weather chart , so keep a record of the weather chart on hourly basis.

Templates : Daily Activity Report Civil and Structural, Daily Activity Report Electrical, Weekly Progress Report, Monthly Progress Report, Weather Chart.

QUALIFIED SELLERS LIST

Is a type of list derived from conducting procurements process. The sellers that are included in the list are already prequalified or prescreened from previous project performance, company, or organization capability. This document is often done during the third step of managing construction projects.

Template : Qualified Sellers List

SELLERS PROPOSAL

A seller proposal is a document submitted by the sellers in response to the procurement documents given to them. This will be the basic information that the evaluation body will be used in selecting qualified sellers. This document is usually done during the third step of managing construction projects.

Template : Sellers Proposal , Tender Form,

BIDDER CONFERENCES

Bidder conferences, pre-bid conferences, and contractor conferences are formal meetings that are conducted prior to submission of proposals or bids. The meeting is often attended by the buyer or owner, project management team, and the prospective sellers. This meeting is usually held during the third step of managing construction projects.

Template : Selected Seller's List

PROCUREMENT CONTRACT

Is a mutually binding legal document or agreement where the seller provides specified requirements, products, or services, and in which the buyer has to compensate the seller. This is in the form of simple or complex document, or just a simple purchase order (P.O.) form. The procurement contract is to be awarded to selected seller.

This document is usually done during the third step of managing construction projects.

Template: Resource Calendar Materials, Resource Calendar Equipment.

	Company Address :
	Telephone :
Your Logo here!	Fax :
	Email : info@
	Website : www.

Work Performance Information

		Date Prepared : 02/27/15	Template ID: ET-I-001
Project	Name of the Project	Revision : 000	Pages : 1 of 1
Project Package	Name of the package / phase of the project		
Project Location	Name of the location of the project	Reference No. : CM-PEC-D-0041-2015	

ID	Category	Deliverables	Activity	Schedule		Cost	Supplier / Contractor	Reference	Status
				Started	Finished	Amount			

Your Logo Here!

Concrete Pouring

Company Address :
Telephone :
Fax :
Email : info@
Website : www.

Date Prepared : 09/01/13	Template ID: ET-Q-001
Revision : 000	Pages : 1 of 1
Reference No.: CM-PEC-D-0042-2013	

Project	Name of the Project
Project Package	Name of the package / phase of the project
Project Location	Name of the location of the project

Time	Truck No.	Slump		Temperature		Concrete Class	Structure Location and Description	Concrete Cylinder ID	Comments
		Required	Reading	Required	Reading				

Contractor's Quality Control :

Signature : _____ /Date
Name :
Organization :

Verified / Checked :

Name / Signature: _____ /Date
Inspector/ Consultant

Noted: _____ /Date
Project Manager

	Company Address :
	Telephone :
	Fax :
	Email : info@
	Website : www.

Concrete Cylinder Test Log

		Date Prepared : 09/01/13	Template ID: ET-Q-002
Project	Name of the Project	Revision : 000	Pages : 1 of 1
Project Package	Name of the package / phase of the project	Reference No. : CM-PEC-D-0043-2013	
Project Location	Name of the location of the project		

Cylinder ID	Location / Member	Date of Pouring	Date of Testing	Age (Days)	Strength(psi) @ 28 days		Test Results		Remarks
					Strength	Reference	Strength	Reference	

Contractor's Quality Control :

Signature : _____ /Date

Name :

Organization :

Verified / Checked :

Name / Signature: _____ /Date

Inspector/ Consultant

Noted: _____ /Date

Project Manager

Template ID : ET-Q-003
Revision - 00
Page 1 of 1

Reference No.: CM-PEC-D-0044-2013

STEEL REINFORCEMENT TEST LOG

Project :
Project Package :   Date :

Item No.	Sample Description	Location / Member	Date Submitted	Date of Testing	Test in TENSION			Test in BENDING		Remarks
					Test Results - Fy (Mpa)		Required Strength - Fy (Mpa)	Test Results	Standard Specs.	
					sample 1	sample 2 average		Bending Angle 180 degrees		
1	28mm Grade 60 - W		06/02/12	06/02/12	447	447	414	No Crack	Non - Crack	satisfactory

prepared by: checked / verified: Noted by:

Name   **Name**   **Name**

Contractor's Quality Control   Inspector /Consultant   Project Manager

Batch Plant Inspection

		Company Address :
		Telephone :
		Fax :
		Email : info@
		Website : www.

		Date Prepared : 09/01/13	Template ID: ET-Q-004
Project	Name of the Project	Revision : 000	Pages : 1 of 1
Project Package	Name of the package / phase of the project		
Project Location	Name of the location of the project	Reference No. : CM-PEC-D-0045-2013	

No.	Category	Rating	Comments
1	Coarse Aggregate Stockpile		
2	Fine Aggregate Stockpile		
3	Cement Warehouse / Trucks		
4	Cement Silo		
5	Transit Mixers		
6	Admixture warehouse / facilities		
7	Source of Water		
8	Water storage tanks		
9	Material Receiving area		
10	Offices / Service Area		
11	Housekeeping		
12	Plant Capability and Performance		

Attached Plant Photographs:

Rating :

    Excellent : 4
    Good Condition : 3
    Fair : 2
    Poor : 1

Inspected by:

_____

Name / Signature

Role

Organization

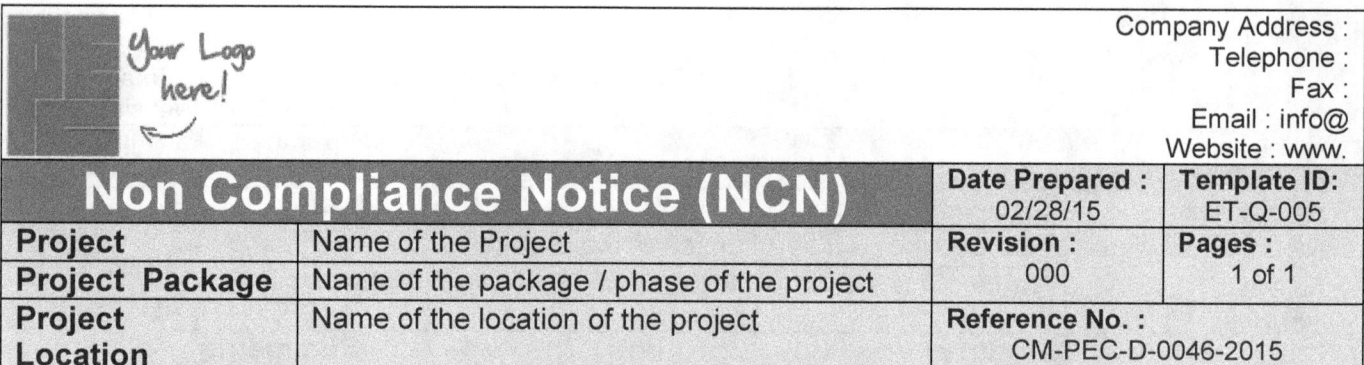

		Company Address : Telephone : Fax : Email : info@ Website : www.	
Non Compliance Notice (NCN)		**Date Prepared :** 02/28/15	**Template ID:** ET-Q-005
Project	Name of the Project	**Revision :** 000	**Pages :** 1 of 1
Project Package	Name of the package / phase of the project		
Project Location	Name of the location of the project	**Reference No. :** CM-PEC-D-0046-2015	

Contractor's/Seller's Company Name Company A		**Contractor's/Seller's Name : In charge** Engineer X **Role :** Construction Manager
WBS ID :	**Scope / Work Package / Requirement:** Retaining Wall	**Reference :** (Drawing , Specification , Contract provision etc.)

No.	Description	Structure location /Particular	Current Status / Condition
1	Spacing of vertical Reinforcement	Retaining Wall at Grid line 1 and Grid line D-G	Installation of reinforcement steel bar

Issued by ; _____ QC / Inspector **Noted by ;** _____ Project Manager	**Received by ;** _____ Contractor / Seller **Date Received ;** _____

Pier Engineering & Consultants
Project Managers • Engineers • Construction

Company Address:
Tel :0000000
Fax : 0000000
Email : info@
Website: www.

Non Compliance Action Form

| Date Prepared : 02/28/15 | Template ID: ET-Q-006 |

Project	Name of the Project	Revision : 000	Pages : 1 of 1
Project Package	Name of the package / phase of the project		
Project Location	Name of the location of the project	Reference No. : CM-PEC-D-0047-2015	

Contractor's/Seller's Company Name Company A	Contractor's/Seller's Name : In charge Engineer X Role :Construction Manager
WBS ID : **Scope / Work Package / Requirement:** Retaining Wall	**Reference :**(Drawing, Specification, Contract provision etc.)

Corrective Action / Defect Repair

No.	Description	Action	Date of Completion	Inspector / Consultant's Comment
1	Spacing of vertical Reinforcement	Re work and corrected into specified spacing		

Certified by ;

_____ / Date _____
Contractor's Quality Control

Verified by ;

_____ / Date _____
Contractor's Construction Manager

Received / Approved ;

_____ / Date _____
Inspector / Consultant

Noted by ;

_____ / Date _____
Project Manager

Your Logo here!			Company Address : Telephone : Fax : Email : info@ Website : www.	
Change Request Form			Date Prepared : 02/28/15	Template ID: ET-Q-007
Project	Name of the Project		Revision : 000	Pages : 1 of 2
Project Package	Name of the package / phase of the project			
Project Location	Name of the location of the project		Reference No. : CM-PEC-D-0048-2015	
Change Number 01	**Change Category**			
	[ ] Scope [ ] Quality [ ] Schedule [ ] Cost [ ] Documents			

Requesting Person			
Name	Role	Organization	Signature

Proposed Change	
Description	Justification

Impact	
Scope	Enumerate and describe the impact of change on the project scope
Quality	Enumerate and describe the impact of change on the project quality
Schedule	Enumerate and describe the impact of change on the project schedule
Cost	Enumerate and describe the impact of change on the project cost

Attachments / Supporting Documents

Evaluator or Reviewed by:		
_____ Name & Signature _____ Role & Organization	[ ] Submit detailed estimates On _____ [ ] Submit Drawings [ ] Attach other documents. **Other Comments ;**	[ ] Approved and proceed to Change Order Authorization [ ] Rejected **References ;** Contracts Drawings Specification , etc.

		Company Address :
		Telephone :
		Fax :
		Email : info@
		Website : www.

Change Request Form

Date Issue : 02/28/15	Pages : 2 of 2
Reference No.: CM-PEC-D-0048-2015	
Revision No. : 000	

Change Control Board				
Name	Role	Organization	Signature	Date

www.ConstructionProjectManagementPro.com

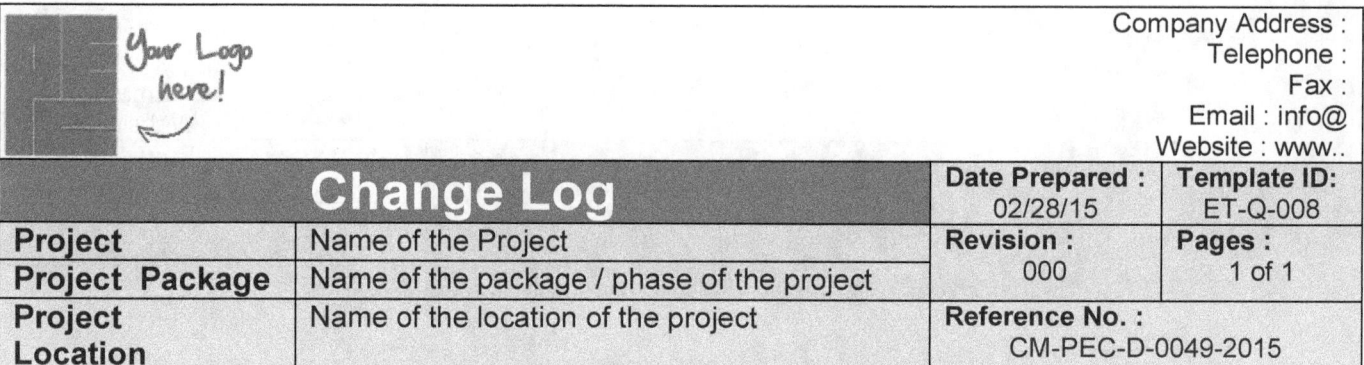

	Change Log		Date Prepared: 02/28/15	Template ID: ET-Q-008
Project	Name of the Project		Revision: 000	Pages: 1 of 1
Project Package	Name of the package / phase of the project			
Project Location	Name of the location of the project		Reference No.: CM-PEC-D-0049-2015	

Change Request No.	Change Description	Submitted		Status			Comments
		Date	Responsible	Approve	Reject	Date	

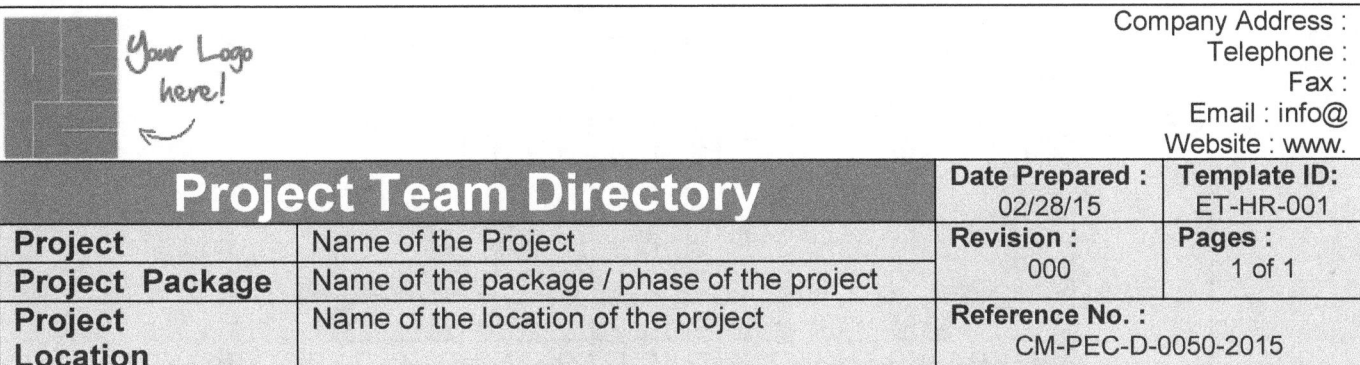

			Project Team Directory		Date Prepared: 02/28/15	Template ID: ET-HR-001
Project		Name of the Project			Revision: 000	Pages: 1 of 1
Project Package		Name of the package / phase of the project				
Project Location		Name of the location of the project			Reference No.: CM-PEC-D-0050-2015	

Name	Role	Address	Contact Numbers			Email
			Telephone	Mobile	Fax	

		Company Address :
		Telephone :
Your Logo here!		Fax :
		Email : info@
		Website : www.

Resource Calendar - Manpower

		Date Prepared : 02/28/15	Template ID: ET-HR-002
Project	Name of the Project	Revision : 000	Pages : 1 of 1
Project Package	Name of the package / phase of the project (*Bored Piling Works*)		
Project Location	Name of the location of the project	Reference No.: CM-PEC-D-0051-2015	

Item No.	Role	weeks											
		1	2	3	4	5	6	7	8	9	10	11	12
1	Project Manager												
2	Project Engineer / Resident Engineer												
3	Material Engineer												
4	Safety Officer												
5	Health Personnel												
6	Foreman												
7	Mason												
8	Carpenter												
9	Labourer												
10	Rigger												
11	Driver												
12	Operator												

This illustration is for illustrative only and not intended to portray any construction projects, scope, description of any type of construction projects.

www.ConstructionProjectManagementPro.com

		Company Address : Telephone : Fax : Email : info@ Website : www.	
Transmittal Form		Date Prepared: 02/28/15	Template ID: ET-COM-001
Project	Name of the Project	Revision : 000	Pages : 1 of 1
Project Package	Name of the package / phase of the project		
Project Location	Name of the location of the project	Reference No. : CM-PEC-D-0052-2015	

TO :
ATTENTION :
CC :

FROM :
ROLE :
ORGANIZATION : **CONTACT #** :

- [ ] FOR YOUR INFORMATION
- [ ] FOR YOUR ACTION
- [ ] FOR YOUR APPROVAL
- [ ] FOR COORDINATION
- [ ] FOR YOURETENTION
- [ ] FOR YOUR COMMENTS/CHECKING & RETURN
- [ ] FOR YOUR SIGNATURE & RETURN
- [ ] FOR YOUR QUOTATION
- [ ] EXTRA COPIES AS REQUESTED
- [ ] AS REQUESTED
- [ ] CHARGEABLE

Sent Via			
[ ] Courier / Mail	[ ] Email /Internet	[ ] Hand	[ ] Self Collect

Number of Prints			Serial No.	Revision	Description
Paper	Tracing Paper	Size			

Remarks :

Acknowledgement			
I / We received the above in good order except serial number **0** as marked "x"			
Company Stamp	Name of Receiver	Signature	Date

www.ConstructionProjectManagementPro.com

		Company Address :
Your Logo here!		Telephone :
		Fax :
		Email : info@
		Website : www.

Project Records Registry – IN		**Date Prepared:** 02/28/15	**Template ID:** ET-COM-002
Project	Name of the Project	**Revision :** 000	**Pages :** 1 of 1
Project Package	Name of the package / phase of the project		
Project Location	Name of the location of the project	**Reference No. :** CM-PEC-D-0053-2015	

No.	Reference No.	Category	Description	Date Received	Comments

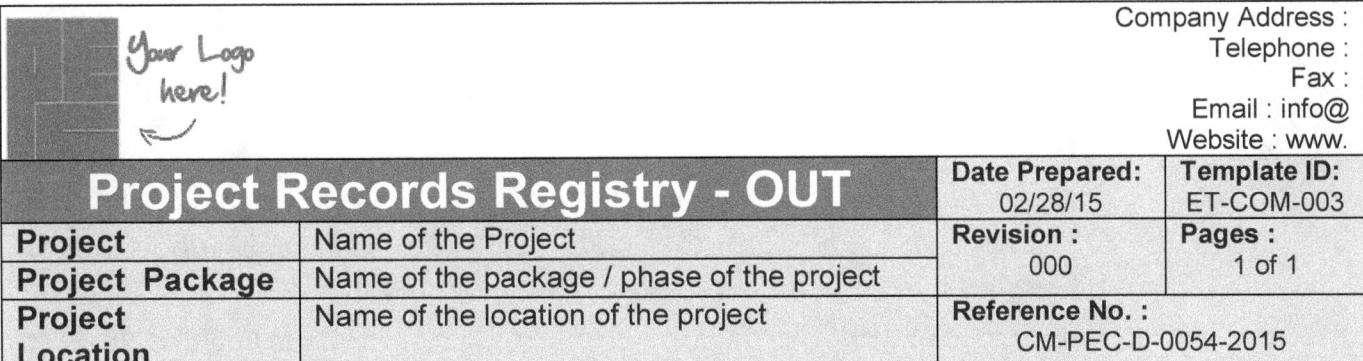

		Company Address :
		Telephone :
		Fax :
		Email : info@
		Website : www.

Project Records Registry - OUT		**Date Prepared:** 02/28/15	**Template ID:** ET-COM-003
Project	Name of the Project	**Revision :** 000	**Pages :** 1 of 1
Project Package	Name of the package / phase of the project		
Project Location	Name of the location of the project	**Reference No. :** CM-PEC-D-0054-2015	

No.	Reference No.	Category	Description	Date Forwarded	Comments

www.ConstructionProjectManagementPro.com

Company Address:
Tel :0000000
Fax : 0000000
Email : info@
Website: www

Template ID: ET-COM-004
Revision: 000

Reference No.: CM-PEC-L-0001-2015

Date: February 28, 2015
Page 1 of 1

CONTRACTOR'S COMPANY NAME
Address

Attention   :CONTRACTOR'S NAME/COMPANY REPRESENTATIVE
                     Role

Project   :**NAME OF THE PROJECT**

Project Package   : **GENERAL CONSTRUCTION WORKS**

Location   :Project Location

Subject   :**SUB- PUMPS / PROTECTION COVER AND SITEINCHARGE DURING HOLY DAYS**

Hi _____ ,

As part of preventive action for possible rain (unpredictable) damage during Holydays , please provide the following as needed on site ;

1. Protective Cover (blue sheet) for sloped area as shown in the picture.
2. Sub-pump
3. Staff / Personnel to check the site condition.

All of the above mention requirements were discussed on site together with Engr._____.

For Your compliance,

Very truly yours,

Name
Consultant's company name

Your LOGO Here!

Company Address:
Tel :0000000
Fax : 0000000
Email : info@
Website: www

Template ID: ET-COM-005
Revision: 000

Reference No. : CM-PEC-L-0002-2015

Date: February 28, 2015

Project : Name of the project	Field Instruction Number :
Project Package : PP 01 – General Civil and Architectural Works	Date : Feb 15, 2015
Location :	
SUBJECT :	
Instructions :	
Remarks :	

Prepared by :

_____

Signature

_____

Printed Name / Date

_____

Role (Field Inspector)/ Organization

Noted by:

_____

Project Manager

www.ConstructionProjectManagementPro.com

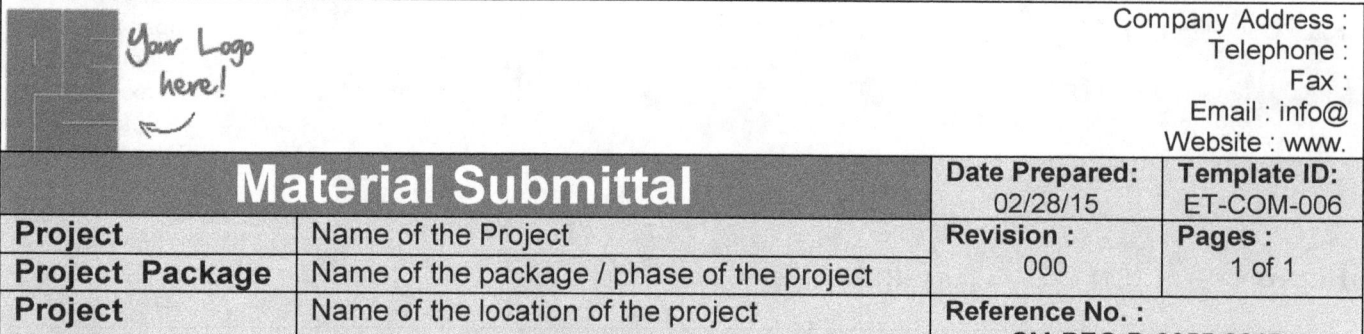

			Company Address : Telephone : Fax : Email : info@ Website : www.	
Material Submittal			**Date Prepared:** 02/28/15	**Template ID:** ET-COM-006
Project	Name of the Project		**Revision :** 000	**Pages :** 1 of 1
Project Package	Name of the package / phase of the project			
Project Location	Name of the location of the project		**Reference No. :** CM-PEC-D-0055-2015	

Category : Civil / Structural			Work Package : Blind Drain		
Item No.	Material Description	Brand Model (Technical Data)	Action Code	Action Date	Remarks
1	GeoTextile	B-20 GeoTextile			

Having reviewed this submittal , we certify that it conforms to requirements and condition of the contract

_____
Contractor / Seller

Project Manager's Comment ;

_____
Signature

_____
Name and Organization

Action Code :
A - Approved , **B** - Approved with comments
C - Revise and Re-submit , **D** - Disapproved

Engineer's /Architect's Comment ;

_____
Signature

_____
Name and Organization

Sponsor's / Customer's Comment ;

_____
Signature

_____
Name and Organization

Template ID : ET-COM-007
Revision : 00
Page 1 of 1

Reference No.: CM-PEC-D-0056-2015

MATERIAL SUBMITTAL SUMMARY

Project :
Project Package :

Date :

No.	Control No.	Revision	Submittal Subject	PLANNED (Contractor's Submission)	ACTUAL (Contractor's Submission)	Consultant's Comment	Approval Status
1	S-M / 0064	0	B-20 Geotextile - Brand Name/ Distributors Name		05/22/15	06/06/15	B
2	S-M / 0001	0	Soil Poisoning Chemicals - Brand Name/ Distributors Name		05/18/15	06/06/15	B
3	S-M 0077/0082	1	Panel Boards & BusDuct - Brand Name/ Distributors Name		06/26/15	06/26/15	B
4							
5							
6							
7							
8							
9							
10							
11							

ACTION CODE / STATUS :: A- Approved , B- Approved with Comments , C- Revise and Re-submit (Work may NOT PROCEED) , D- Disapproved

prepared by:

Template ID : ET-COM-008
Revision : 00
Page 1 of 1

Reference No.: CM-PEC-D-0057-2015

DRAWING REGISTRY

Project :
Project Package :
Date :

Item No.	WBS ID	Description	Sheet Number	Revision Number	Revision Date	Consultant (Project Management) Date received from designer	Consultant (Project Management) Date issued to Contractor	Size and Type	Purpose	Remarks
1		BORED PILES								
		Bored pile lay out	BP-0010					A1 (blueprint)	for bidding	
		Pile cap details	BP-0011					A1 (blueprint)	for bidding	
			BP-0012					A3 (pdf)	for bidding	

prepared by:

Your LOGO Here!

Company Address:
Tel :0000000
Fax : 0000000
Email : info@
Website: www

Template ID: ET-COM-009
Revision: 000

Reference No.: CM-PEC-L-0003-2015

Date: February 28, 2015

MEMO

TO	:	**ALLCONTRACTORS, SUB CONTRACTORS**
FROM	:	**NAME** **PROJECT MANAGER** ORGANIZATION
SUBJECT	:	**YOUR SUBJECT HERE**

Body of text here.

Respectfully Yours,

Signature:

_____
Project Manager Name

_____
Organization or Company Name

www.ConstructionProjectManagementPro.com

Company Address:
Tel :0000000
Fax : 0000000
Email : info@
Website: www

Template ID: ET-COM-010
Revision: 000

Reference No.: CM-PEC-D-0058-2015   Date: February 28, 2015

Page 1 of 3

Project   :   **NAME OF THE PROJECT**
Location   :   **PROJECT LOCATION**
Subject   :   **PRE CONSTRUCTION MEETING AGENDA**
Venue   :   **LOCATION OF THE MEETING**   Time :

1.0 Project Organization
    1.1   Introduction of the project team
        1.1.1   Owner
        1.1.2   Contractor
            1.1.2.1 CV of Personnel
            1.1.2.2 Site In Charge (in the absence of Contractor's CM)
            1.1.2.3 Organizational Chart
        1.1.3   Designers/Consultants
        1.1.4   Project Management Consultants

2.0 Requirement for Building Permit (by owner or Contractor)
    2.1   Permit Plans *(Sets , sealed)*
    2.2   Permit Forms
    2.3   Environmental Compliance Certificate

3.0 Project Implementation Schedule
    3.1   Bar Chart with S-Curve
    3.2   PERT – CPM
        3.2.1   For Contractor's Planner/Scheduler

4.0 Construction Drawings and Specifications
    4.1   Issuance of final construction drawings and specifications
        4.1.1   Owner
        4.1.2   Contractor
        4.1.3   Project Manager

5.0 Construction Management System
    5.1   Communication System
        5.1.1   Communication Flow
        5.1.2   Email and File Sharing System
    5.2   Construction Standard Templates

Reference No. : CM-PEC-D-0058-2015

Template ID: ET-COM-010
Revision: 000
Date: February 28, 2015

    5.3   Approval turn around period for technical submittal

    5.4   Quality Control Program
            5.4.1   Design Mix
                  5.4.1.1 Batching Plant (with options/alternatives)
            5.4.2   Testing Laboratories *(QC Personnel)*

    5.5   Environment , Health and Safety
            5.5.1   Regular Toolbox Meetings
            5.5.2   Hazard / Risk Identification and Mitigation
            5.5.3   Safety Officer - Full time
            5.5.4   Project Nurse - Full time

6.0 Construction Policies, Requirements

    6.1   Site Access
            6.1.1   Gate and Pathway
            6.1.2   Coordination with Campus Security
                  6.1.2.1 Security Agency
            6.1.3   Worker's ID/PPE/Uniform

    6.2   Site Offices / Warehouse / Temporary Facilities
            6.2.1   Site Lay out
            6.2.2   PM site office requirement
            6.2.3   Site Status as of date

    6.3   Construction Site Security

    6.4   Material and or debris storage and disposal

    6.5   Protection of existing structures / facilities

    6.6   Campus Rules and Regulations
            6.6.1   Deliveries and Daily Access *(coordinate with owner's representative)*
            6.6.2   Working Time *(coordinate with the owner)*
            6.6.3   Other restrictions
            6.6.3.1 No Smoking
            6.6.3.2 Others
    6.7   Construction Methodology
    6.8   Schedule of Regular Construction Coordination Meetings

YOUR LOGO HERE!

Company Address:
Tel :0000000
Fax : 0000000
Email : info@
Website: www

Template ID :ET-COM-010
Revision : 000
Date: February 28, 2015

Reference No. : CM-PEC-D-0058-2015

7.0 Other Matters
    7.1   Ground Rules
        7.1.1 Meetings
            7.1.1.1 Switch off the Cell phone .
        7.1.2 Others
    7.2   Delivery of Materials (contractors responsibility)
    7.3   Temporary Utilities
    7.4   Full Time Basis - Site Personnel
        7.4.1   Civil and Structural Engineer
        7.4.2   Building Services Engineer
        7.4.3   Liaison Officer
        7.4.4   Safety Officer
        7.4.5   Planner
        7.4.6   QA/QC
        7.4.7   Survey Team
        7.4.8   QS

Prepared by:

_____
Name and Signature

Noted by:

_____
Project Manager

Company Address:
Tel :0000000
Fax : 0000000
Email : info@
Website: www

Template ID: ET-COM-011
Revision: 000

Reference No. : CM-PEC-D-0059-2015

Date: February 28, 2015
Page 1 of 3

Project	:	**NAME OF THE PROJECT**
Location	:	**PROJECT LOCATION**
Subject	:	**PRE CONSTRUCTION MINUTES OF MEETING**
Venue	:	**LOCATION OF THE MEETING**

ATTENDEES:

Name	-	Company Name	Name	-	Company Name
Name	-	Company Name	Name	-	Company Name
Name	-	Company Name	Name	-	Company Name

ACCOMPLISHMENT:

WBS ID	Work Package No.	Target Accomplishment	Actual Accomplishment	Slippage/ Advance	Equivalent Days
	Bored Piling				
	Architectural				

The meeting was called to order at TIME and the following items were discussed:

ITEM NO.	MATTERS DISCUSSED	RESPON-SIBLE	DUE DATE	REMARKS
1.0	**Project Organization**			
1.1	**Introduction of the project team**			
1.1.1	**Owner**			
1.1.2	**Contractor**			
1.1.3	**Designers/Consultants**			
1.1.4	**Project Management Consultant** and others will be mobilize upon construction on-progress			
2.0	**Requirement for Permits (by Owner)**			
2.1	**Building Permit**			

YOUR LOGO HERE!

Company Address:
Tel :0000000
Fax : 0000000
Email : info@
Website: www

Reference No. : CM-PEC-D-0059-2015

Template ID :ET-COM-011
Revision : 000
Date : February 28, 2015

2.2	ECC				
3.0	**Project Implementation Schedule**				
3.1	Bar Chart & S-curve				
3.2	PERT CPM				
4.0	**Construction Drawings & Specifications**				
4.1	For Construction Drawings &				
5.0	**Construction Management System**				
5.1	Communication System				
5.2	Construction Standard Templates				
5.3	Approval turn around period for technical submittal.				
5.4	Quality control program				
5.5	Environment, health & Safety				

YOUR LOGO HERE!

Company Address:
Tel :0000000
Fax : 0000000
Email : info@
Website: www

Template ID :ET-COM-011
Revision : 000

Reference No. : CM-PEC-D-0059-2015

Date : February 28, 2015

There being no other matters to be discussed the meeting was adjourned at TIME.

The Minutes of Meeting is the recollection of what transpired at the meeting. If no corrections/amendments or other information within two (2) days from the receipt of the minutes, this will be considered reflective of the accounts of what transpired and therefore considered approved.

Prepared by:

_____
Name and Signature

Noted by:

_____
Project Manager

				Company Address :
				Telephone :
	Your Logo here!			Fax :
				Email : info@
				Website : www.

Daily Activity Report

			Date Prepared:	Template ID:
Day : Monday	Weather : Rainy	Temperature : 20 C	02/28/15	ET-COM-012
Project	Name of the Project		**Revision:**	**Pages :**
Project Package	Name of the package / phase of the project		000	1 of 1
Project Location	Name of the location of the project		**Reference No. :** CM-PEC-REP-0001-2015	
Category	*Write the category of the trade – Civil or Structural*			

ACTIVITIES

ID	Category	Description	Accomplished	Variance	Comments

RESOURCES

ID	Manpower / People	Comments

ID	Materials	Comments

ID	Equipment	Comments

BASELINE

Quality		Cost / Funds		
Variance	Actions	Planned to spent	Actual spent	Variance and Comments

ISSUES :
RISKS :

Prepared by : (Name) _____
　　　　　　　__ (Organization) _____
　　　　　　　__ (Role) _____

www.ConstructionProjectManagementPro.com

Your Logo here!

Company Address :
Telephone :
Fax :
Email : info@
Website : www.

Daily Activity Report

			Date Prepared: 02/28/15	Template ID: ET-COM-013
Day : Monday	Weather : Rainy	Temperature : 20 C		
Project	Name of the Project		Revision : 000	Pages : 1 of 1
Project Package	Name of the package / phase of the project			
Project Location	Name of the location of the project		Reference No. : CM-PEC-R-0002-2015	
Category	Electrical Works			

ACTIVITIES

ID	Category	Description	Accomplished	Variance	Comments

RESOURCES

ID	Manpower / People	Comments

ID	Materials	Comments

ID	Equipment	Comments

BASELINE

Quality		Cost / Funds		
Variance	Actions	Planned to spent	Actual spent	Variance and Comments

ISSUES :

RISKS :

Prepared by : __(Name)_____
__(Organization)_____
__(Role)_____

Project Name

Perspective View

PROJECT WEEKLY UPDATE
Date:

Company Address:
Tel :0000000
Fax : 0000000
Email : info@
Website: www

Template ID : ET-COM-017
Revision : 000

Reference No. : CM-PEC-R-0006-2015

Date : February 28 , 2015

A. Accomplishment Update

WBS ID	Work Package / Description	Planned %	Actual %	Variance	Status as of date (current activity)	Comments
	Bored Pile	9.94	0.13	-9.81	Fabrication of Rebar Boring of pile holes	Schedule: 19 Calendar days- delayed Cost :
	Structural Steel					Schedule: Cost :
	Architectural					Schedule: Cost :
	Electrical					Schedule: Cost :
	Other Items					Schedule: Cost :

Reference No.: CM-PEC-R-0006-2015
Date : February 28 , 2015

B.   Project Update

Item No.	Description	Comments
1	Safety Issues	
2	Design	
3	Procurement	
4	Requirements	

C.   Progress Photos

Description

Reference No. : CM-PEC-R-0006-2015

Date : February 28 , 2015

Site Picture

Description

Site Picture

Description

Project Name
Project Location

Perspective View

PROGRESS REPORT NO. 01
Covering Period of *Month – Month*, 2013

Company Address:
Tel :0000000
Fax : 0000000
Email : info@
Website: www

Template ID: ET-COM-018
Revision: 000

Reference No. : CM-PEC-R-0007-2015

Date : February 28, 2015

OWNER'S COMPANY NAME
Address

Attention	:	**OWNER'S NAME or COMPANY REPRESENTATIVE** *Role*
Project	:	**NAME OF THE PROJECT** *Project Location*
Subject	:	**MONTHLY STATUS REPORT NO.1**

Sir/Ma'am,

We are submitting herewith the **Monthly Status Report No. 01** for the proposed **Name of the project** covering the period from **Date, Year to Date Year.**

We hope you will find everything in order.

Thank you.

Very truly yours,

NAME
CONSULTANT'S COMPANY NAME-Resident Engineer

Noted by:

NAME
CONSULTANT'S COMPANY NAME-Project Manager

Reference No.: CM-PEC-R-0007-2015   Date : February 28 , 2015

Here are the contents of the Monthly Report but not limited to the following;

    I.   GENERAL SUMMARY

    II. CONSTRUCTION

        A. Accomplishment Update
        B. Work in progress and Milestone

    III. DESIGN UPDATE

    IV. PROCUREMENT

        A. Contractors and Suppliers List
        B. Owner Supplied Materials and Equipment Lists (OSME Lists)

    V. FINANCIAL STATUS

    VI. REQUIREMENTS UPDATE

        A. Permits and Licenses

    VII. ATTACHMENTS

        A. Progress Photos
        B. Schedule
        C. Weather Charts
        D. Minutes of Meeting

Template ID : ET-COM-019
Revision No.000
Page 1 of 1
CM-PEC-R-0008-2013

Project Name :

Project Package / Phase :

Project Location :

12 N — 6 AM — 6 PM — 12 MN

1	2	3	4	5	6	7
8	9	10	11	12	13	14
15	16	17	18	19	20	21
22	23	24	25	26	27	28
29	30	31				

LEGEND:

Fair [ ]   Rainy [ ]

Cloudy [ ]   Stormy [ ]

Recorded by:

Noted by:

Construction Manager

			Company Address : Telephone : Fax : Email : info@ Website : www.	
Qualified Sellers Lists			**Date Prepared :** 02/19/15	**Template ID:** ET-P-001
Project	Name of the Project		**Revision :** 000	**Pages :** 1 of 1
Project Package	Name of the package / phase of the project			
Project Location	Name of the location of the project		**Reference No. :** CM-PEC-D-0060-2015	

Work Package (WP)	Description	Seller/Bidder's Company Name	Seller/Bidder's Contact Details	Contact Person	Remarks
WP-01	General Civil/ Architectural Works	Seller No.1 **Company Name:**	Company Details (Address, Telephone , website) :	Name : Role: Mobile No. Email:	
		Seller No.2 **Company Name:**	Company Details (Address, Telephone , website) :	Name : Role: Mobile No. Email:	
		Seller No.3 **Company Name:**	Company Details (Address, Telephone , website) :	Name : Role: Mobile No. Email:	
WP-02	Bored Piling Works	Seller No.1 **Company Name:**	Company Details (Address, Telephone , website) :	Name : Role: Mobile No. Email:	
		Seller No.2 **Company Name:**	Company Details (Address, Telephone , website) :	Name : Role: Mobile No. Email:	

SELLER'S LOGO HERE		Company Address : Telephone : Fax : Email : info@ Website : www.	
Seller Proposal **(Water Proofing Proposal)**		Date Prepared: 02/28/15	Template ID: ET-P-002
Project Package	*Water Proofing*	Revision : 000	Pages : 1 of 2
Project	Name of the project		
Project Location	Name of the location of the project	Reference No. : CM-PEC-D-0061-2015	

OWNER REPRESENTATIVE
Address

Attention   :   **OWNER'S NAME or REPRESENTATIVE**
                              Role

Dear Sir/Ma'am,

In compliance with your request for proposal on the above project package, we are pleased to submit the following for your consideration:

Item No.	Location	Area (sq.m)	Specification	Unit Price / sq.m.	Amount ($)
1	**Concrete Slab**	10,000	(Cementitous/Capillary) **B15** Slurry		
			Total		$

Amount in Words:

Terms of Payment : 30% Down Payment , balance thru weekly progress billing payable within seven (7) days upon receipt of each billing.

Guarantee   :   **B15** Slurry - Waterproofing system against leak for a period of Five (5 years).

Scope of work   :   **OSMOSEAL (Osmosis/ Crystallization)**

1. Mobilization
2. Scrape/ Remove existing loose mortar, dirt and other foreign matters.

Reference No.: CM-PEC-D-0062-2015 Date : February 28 , 2015

3. Construction joints and perimeter of drain pipes with polyurethane sealant .cracks should be out to 1/4" x 1/4" prior to filling with polyurethane sealant. Allow to dry.

4. Saturate routed area and leave damp for application.

5. Apply two (2) coating of *B15* Slurry Cementitous Crystallization Waterproofing on the entire prepared area.

Should you have any queries, please do not hesitate to call us and we will be glad to discuss this with you at your most convenient time.

Very truly yours,

_____
SELLER'S NAME AND SIGNATURE
Role
SELLER'S COMPANY NAME

CONFORME:

Signature :
By   : Owner's Name or Representative
Date   :

		Company Address : Telephone : Fax : Email : info@ Website : www.	
Tender Form		**Date Prepared:** 02/28/15	**Template ID:** ET-P-003
Project	Name of the Project	**Revision :** 000	**Pages :** 1 of 2
Project Package	Fire Protection, Plumbing, HVAC, Electrical, Testing and Commissioning		
Project Location	Name of the location of the project	**Reference No. :** CM-PEC-D-0062-2015	

1. I/We, having read and examined the tender documents and drawings, do hereby offer to execute and complete, in accordance with the Conditions of Contract and other Contract Documents, the whole of the *Project Package* as follows -

   For completion of the whole of The Works within your set time as stated in your Conditions of Contract (Appendix A) for the sum of $

   ....

   ($) including all Taxes.

2. I/We agree that should any discrepancy occur between the amounts written in words and in figures entered upon the Form of Tender, the amount written in words will be used.

3. I/We agree to abide by this tender for the period of 90 days from the date fixed for receiving the same and shall remain binding upon me/us and maybe accepted at any time before the expiration of that period.

4. If my/ our tender is accepted, I/we will provide you with a performance bond, callable upon demand, either in cash or by way of an approved insurance bond or banker's guarantee of an amount equal to 20% of the Contract Sum within 7 days of the date of acceptance of my/our tender.

5. I/We agree that should any arithmetical or obvious pricing errors be discovered before acceptance of this offer in the priced tender documents submitted by me/us, then these errors shall be corrected in order that the corrected total of the Final Summary equates with the amount entered upon the Form of Tender in accordance with the following procedure:

   The net total of such errors, whether addition or net omission, will be calculated as a percentage of the corrected total of the Final Summary and all unit rates and prices throughout the Bills of Quantities shall be considered as reduced or increased, as the case may be, by such percentage.

Reference No. : CM-PEC-D-0061-2015 Date: February 28, 2015

6. I/We undertake, in the event of your acceptance, to execute with you a formal contract embodying all the conditions and terms contained in this offer and until such execution, the conditions and terms contained in this offer shall bind me/us and your acceptance shall constitute a binding contract between you and me/us.

7. I/We understand that you are not bound to accept the lowest or any tender.

8. Unless directed to the contrary, I/we undertake to commence The Works immediately but not later than 7 days from issuance of Letter of Award.

9. I/We acknowledge receipt of Tender Addendum/Bulletin No(s) ____ to the Tender Documents and confirm that the contents of the said Tender Addenda/Bulletin form an integral part of the tender submitted by me/us.

10. If there shall be a wage increase during the construction period the total labor adjustment per one USD (1.00) increase in minimum wage or compulsory allowances to be paid by the Owner to the Contractor is a lump sum of USD: _____, which shall be applied to the remaining works after the wage increase and according to the provisions of the Contract Documents.

11. I/We confirm that all bonds, insurances, permits and other fees are to be shouldered by our company unless stated in the Contract Documents to be by Owner. All penalties incurred shall be my/our responsibility.

12. I/We will complete the whole Works within _____ (___) calendar months inclusive of all Sundays and holidays.

Signature :
in the capacity of : *Name of the Owner or Representatives*
Duly authorized to sign tenders on behalf of : *Company Name*
Company Address :
Date Signed :

Witness : *Name of Witness*
Address :
Date Signed :

Please seek advice from your attorney or legal advisor when dealing with this tender form.

Template ID : ET-P-005
Revision No. : 000
Pages : 1 of 1
CM-PEC-D-0063-2015

SELECTED SELLER'S LIST

Project :
Project Package :
Project Location :
Date prepared :

Project Package No.	Description	Bidders / Contractor/Supplier	Cost			Total Amount	Remarks
			Materials	Labor	Equipment		
PP-01	Civil / Architectural Works	Company X					
PP-02	Plumbing and Sanitary	Company X					
PP-03	Mecahnical	Company X					
PP-04	Electrical	Company X					
PP-05	Fire Protection System	Company Y					
PP-06	CCTV Surveillance System	Company Z					
PP-07	Structured Cabling	Company Z					
PP-08	Audio / Video and Equipment	Company A					
PP-09	Fire Alarm and Detection System	Company B					

Prepared by:

Resource Calendar - Materials

Company Address :
Telephone :
Fax :
Email : info@
Website : www.

Template ID:	ET-P-007
Pages :	1 of 1
Date Prepared :	02/28/15
Revision :	000
Reference No. :	CM-PEC-D-0064-2015

Project	Name of the Project
Project Package	Name of the package / phase of the project
Project Location	Name of the location of the project

No.	Description of Material	Make		Required at Site	Planned Submittal Date	1st Submittal Date	RESPONSE	Approval Date		Order Date		Actual Delivery Date at Site	Remarks/Status
		Specified	Proposed					Rcpt.	Actual	Planned	Actual		
1													
2													
3													
4													
5													
6													
7													
8													

STATUS:- A- Approved, B- Approved with Comments; C- Revise and Re-submit , D- Disapproved

Prepared by:

Resource Calendar - Equipment

Company Address :	
Telephone :	
Fax :	
Email : info@	
Website : www.	

Date Prepared : 02/28/15	Template ID: ET-P-008
Revision : 000	Pages : 1 of 1
Reference No. : CM-PEC-D-0065-2015	

Project	Name of the Project
Project Package	Name of the package / phase of the project (*Bored Piling Works*)
Project Location	Name of the location of the project

Item No.	Equipment Description	No. of Units	1	2	3	4	5	6	7	8	9	10	11	12
1	Boring Machine	2		▓	▓	▓	▓	▓	▓	▓	▓			
2	Generator Set	2		▓	▓	▓	▓	▓	▓	▓	▓			
3	Diesel Hammer	4			▓	▓	▓	▓	▓	▓	▓			
4	Crawler Crane – 40 Tonner	2			▓	▓	▓	▓	▓	▓	▓			
5	Feeder Crane – 25 Tonner	2				▓	▓	▓	▓	▓	▓			
6	Welding Machine	2				▓	▓	▓	▓	▓	▓	▓		
7	Concrete Batching Plant	1					▓	▓	▓	▓	▓	▓		
8	Transit Mixer	3					▓	▓	▓	▓	▓	▓		
9	Concrete Vibrator	3					▓	▓	▓	▓	▓	▓		
10	Pay Loader	1						▓	▓	▓	▓	▓	▓	
11	Dump Truck	2						▓	▓	▓	▓	▓	▓	▓
12	Boom Truck	1							▓	▓	▓	▓	▓	

This illustration is for illustrative only and not intended to portray any construction projects, scope, description of any type of construction projects.

Step 4
Monitoring and Controlling

www.ConstructionProjectManagementPro.com

STEP 4 : MONITORING AND CONTROLLING

Monitoring and Controlling is the fourth step in managing construction projects with the purpose of reviewing , monitoring and tracking of project progress and performance and compare it to a given baseline of the project management plan , monitoring the project gives an overview about the project performance and it will identify some areas of the project that needs additional attention , controlling contains preventive or corrective actions and assures that defined actions resolved the project issue .

Flow Chart - Project Management process is not always sequential or performed in identical sequence.

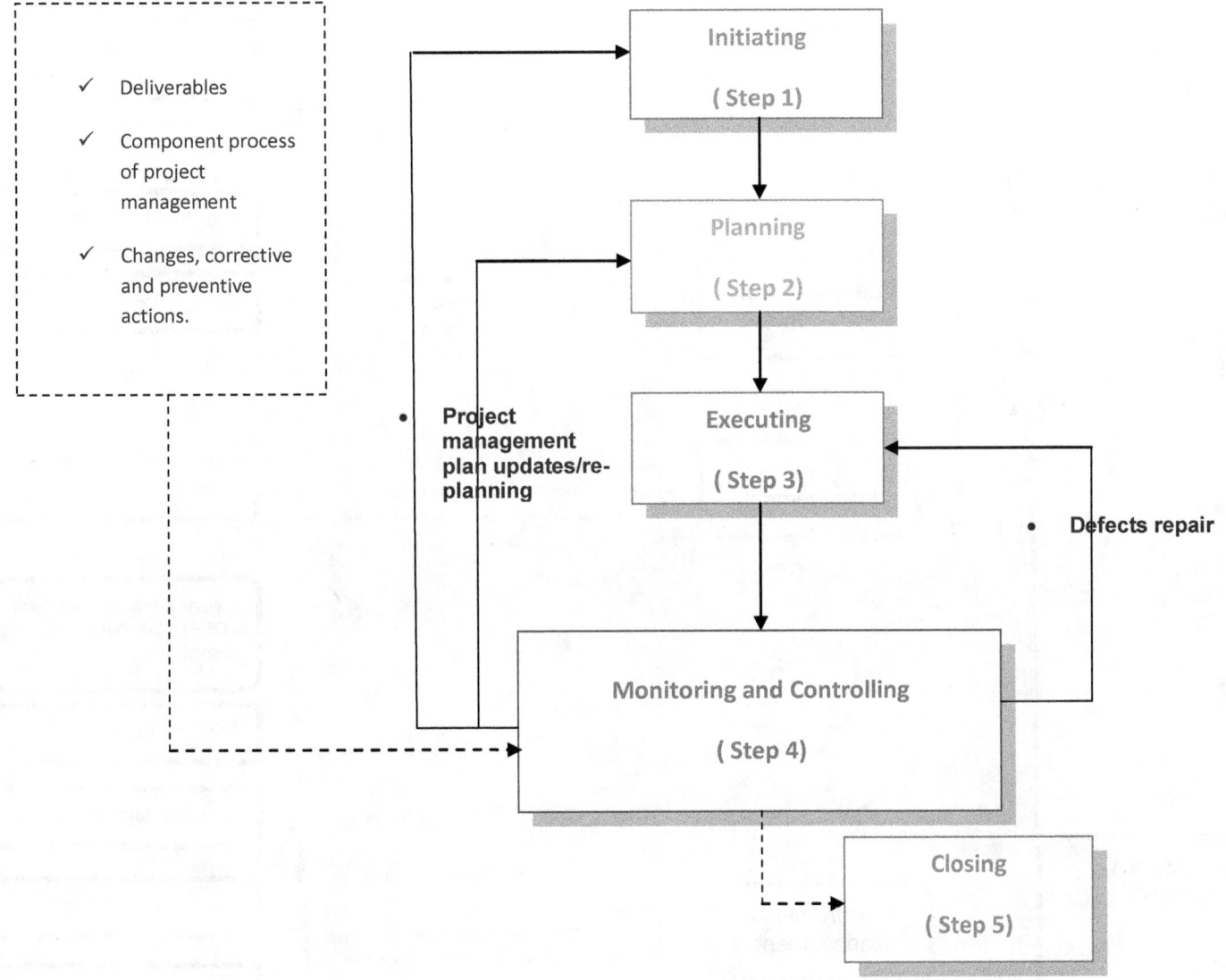

Construction Management: Step-by-Step Templates — Step 4 : Monitoring and Controlling

Mapping:

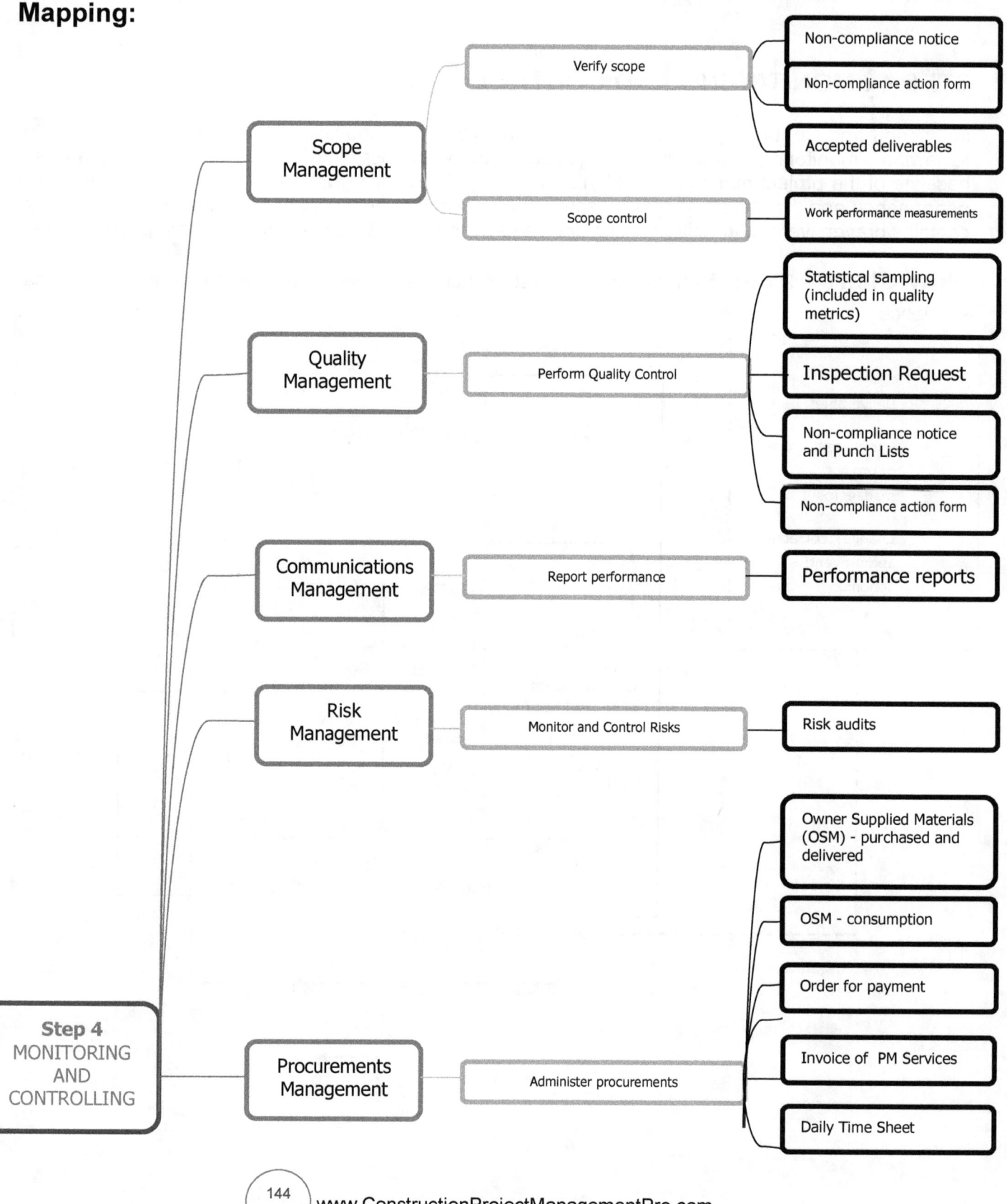

Construction Management: Step-by-Step Templates   Step 4 : Monitoring and Controlling

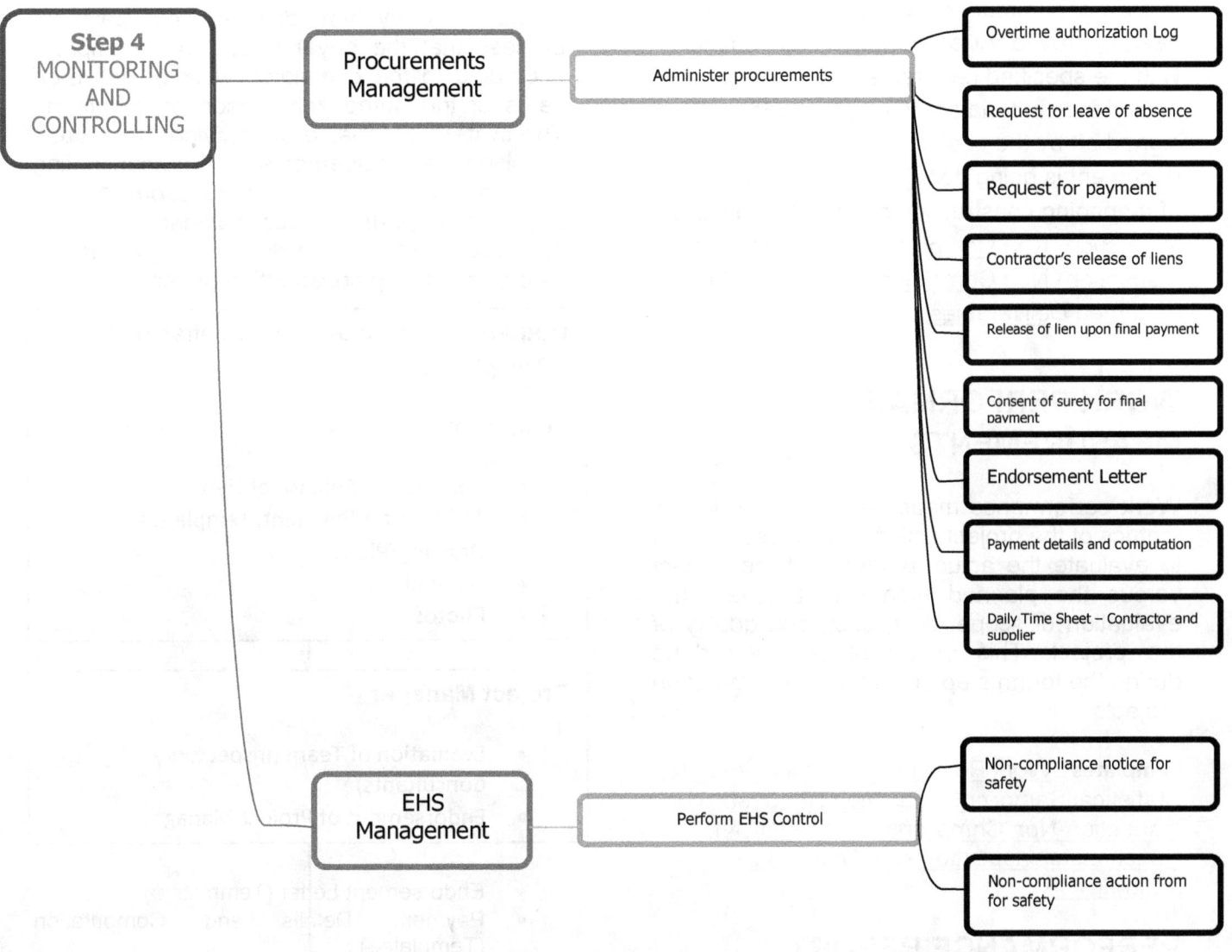

ACCEPTED DELIVERABLES

Is a list or summary of deliverables that has been approved. All deliverables that complied with the specified requirements and/or acceptance criteria are to be approved and signed off by the customer or sponsor. This document is being done during the fourth step of managing construction projects. **Templates** :
Inspection- Non Compliance Notice (NCN),
Inspection- Non Compliance Action (NCA),
Accepted Deliverables

WORK PERFORMANCE MEASUREMENTS

Work performance measurements refer to the metrics of the project activity that is established to evaluate the actual progress of the project versus the planned progress. It covers the evaluation of scope, time, cost, and quality of the project. This document is being done during the fourth step of managing construction projects.

Templates : Work Performance Measurements, Statistical Sampling, Inspection Request, Inspection-Non Compliance Notice (NCN), Punch Lists, Inspection-Non Compliance Action (NCA)

PERFORMANCE REPORT

Performance report illustrates the overall progress or result of the project. It is collected progressively and ought to be sent to stakeholders as part of communication process. The information that needs to be communicated includes the project status, its progress, and performance measurement for cost, schedule, scope, and quality. This document is often done during the fourth step of managing construction projects.
Templates : Performance Report, Risk Audit, Owner Supplied Materials-Purchased, Owner Supplied Materials- Consumption.

PAYMENT SYSTEM

A payment system dictates the payment process that the buyer needs to accomplish, according to the completed work or delivered items of the authorized person on the team. The system is capable of verifying these data, checking the requirements, and authenticating the contract provisions prior to submission to accounts department for payment process. This document is being done during the fourth step of managing construction projects.

Request for Payment from Contractor/ Seller – Template #

Attachments:

- Contractors Release of Lien
- Affidavit for Payment, Template #
- Drawing/Plans
- Schedule
- Photos

Project Manager

- Evaluation of Team (Inspectors / Consultants)
- Endorsement of Project Manager

- Endorsement Letter (Template #)
- Payment Details and Computation (Template#)

Sponsor;

- For Final Approval
- To be forwarded to Accounts payable system
- Payments

Templates : Order For Payment

Templates : Invoice of PM Services, Daily Time Sheet, Overtime Authorization Log, Request For Leave, Request For Payment, Contractor's Release of Lien, Release of Lien upon final payment, Consent of surety for final payment, Endorsement Letter, Payment details and computation, Daily time sheet –seller, Non-compliance Notice (NCN) for Safety, Non-compliance Action (NCA) for Safety.

			Company Address :
Your Logo here!			Telephone :
			Fax :
			Email : info@
			Website : www.

Non Compliance Notice (NCN)

		Date Prepared: 02/28/15	Template ID: MCT-SC-001
Project	Name of the Project	Revision: 000	Pages: 1 of 1
Project Package	Name of the package / phase of the project		
Project Location	Name of the location of the project	Reference No. : CM-PEC-D-0066-2015	

Contractor's/Seller's Company Name		Contractor's/Seller's Name : In charge
Company A		Engineer X
		Role : Construction Manager
WBS ID:	**Scope / Work Package / Requirement:** Retaining Wall	**Reference :** (Drawing , Specification , Contract provision etc.)

No.	Description	Structure location/ Particular	Current Status / Condition
1	**Total number** of vertical Reinforcement	Retaining Wall at Grid line 1 and Grid line D-G	Installation of reinforcement steel bar

Issued by ;
_____
QC / Inspector

Noted by ;
_____
Project Manager

Received by ;
_____
Contractor / Seller

Date Received ;
_____

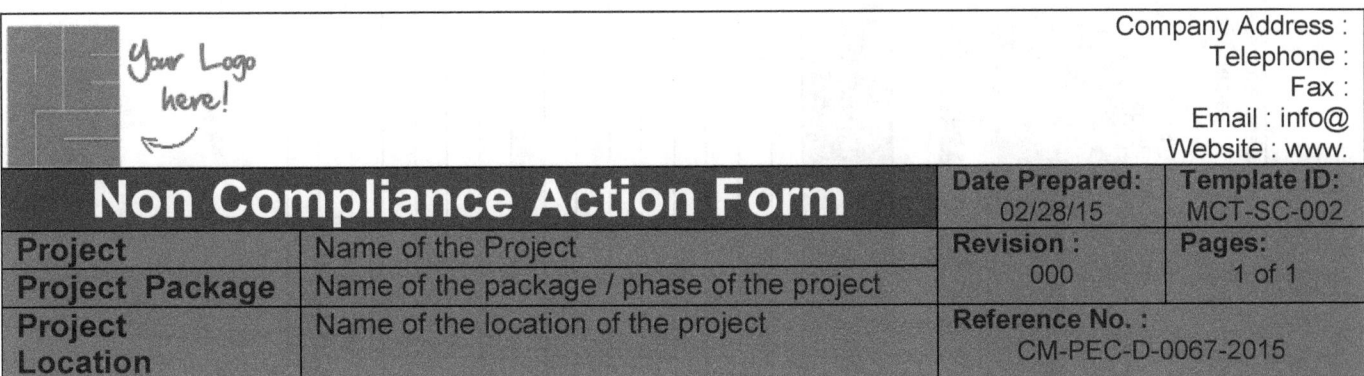

Non Compliance Action Form

Project	Name of the Project	Date Prepared: 02/28/15	Template ID: MCT-SC-002
Project Package	Name of the package / phase of the project	Revision: 000	Pages: 1 of 1
Project Location	Name of the location of the project	Reference No.: CM-PEC-D-0067-2015	

Contractor's/Seller's Company Name Company A	Contractor's/Seller's Name : In charge Engineer X **Role** :Construction Manager
WBS ID : **Scope / Work Package / Requirement:** Retaining Wall	**Reference :**(Drawing , Specification , Contract provision etc.)

Corrective Action / Defect Repair

No.	Description	Action	Date of Completion	Inspector / Consultant's Comment
1	**Total number** of vertical Reinforcement	Re work and corrected as per specified total number of bars		

Certified by ;

_____ / Date _____
Contractor's Quality Control

Verified by ;

_____ / Date _____
Contractor's Construction Manager

Received / Approved ;

_____ / Date _____
Inspector / Consultant

Noted by ;

_____ / Date _____
Project Manager

Accepted Deliverables

Company Address :		
Telephone :		
Fax :		
Email : info@		
Website : www.		

Template ID: MCT-SC-003	Date Prepared : 02/28/15	
Pages : 1 of 1	Revision : 000	
	Reference No. : CM-PEC-D-0068-2015	

Project	Name of the Project
Project Package	Name of the package / phase of the project
Project Location	Name of the location of the project

ID	Category	Deliverables	Requirements			Status	Acceptance
			Specified	Acceptance Criteria	Reference	Verification	

Work Performance Measurements

Company Address :		
Telephone :		Fax :
Email : info@		
Website : www.		

	Date Prepared : 02/28/15	Template ID: MCT-SC-004
Project	Name of the Project	Revision : 000
Project Package	Name of the package / phase of the project	Pages : 1 of 2
Project Location	Name of the location of the project	Reference No. : CM-PEC-D-0069-2015

ID	Category	Deliverable	Schedule		Cost		Scope	Quality
			Value	Interpretation	Value	Interpretation	(scope / technical performance)	(technical performance)
			SV	✓	CV	✓		
			SPI	✓	CPI	✓		
			Schedule		**Cost**		**Scope**	**Quality**
			Value	Interpretation	Value	Interpretation	(scope / technical performance)	(technical performance)
			SV	✓	CV	✓		
			SPI	✓	CPI	✓		

	Company Address :
Your Logo here!	Telephone :
	Fax :

Work Performance Measurements

Date Issue :	02/28/15
Reference No. :	CM-PEC-D-0069-2013
Pages :	2 of 2
Revision No. :	000
Email : info@	
Website : www.	

ID	Category	Deliverable	Schedule		Cost		Quality
			Value	Interpretation	Value	Interpretation	(technical performance)
			SV	✓	CV	✓	
			SPI	✓	CPI	✓	
			Scope (scope / technical performance)				
			Schedule		Cost		Quality
			Value	Interpretation	Value	Interpretation	(technical performance)
			SV	✓	CV	✓	
			SPI	✓	CPI	✓	
			Scope (scope / technical performance)				

Company Address :	
Telephone :	
Fax :	
Email : info@	
Website : www.	

Statistical Sampling

Project	Name of the Project	**Template ID:** MCT-Q-001
Project Package	Fire Protection, Plumbing, HVAC, Electrical, Testing and Commissioning	**Date Prepared :** 02/28/15
		Revision : 000
		Pages : 1 of 1
Project Location	Name of the location of the project	**Reference No. :** CM-PEC-D-0070-2015

ID	Category	Item	Description	Method of Measurements	Metrics	Reference	Statistical Sampling
PP-01	Bored Pile	Concrete	Cast in place concrete pile (bored pile)	Compressive Test	3,000 psi (@28 days)	Structural Plan	• 5 Sets of cylinder at 3 piles / day
				Slump	100mm (max.)	Structural Plan / Structural Specification	• For each batch of concrete • 6 cu.m which ever is lesser
		Reinforcing Steel	Bored pile reinforcing steel	Tensile (Fy) 12mm dia. Bar and larger	413.7Mpa (60 ksi)		2.5 Tons per diameter per kind
				10mm dia bar And smaller	276 Mpa (40 ksi)		2.5 Tons per diameter per kind
				Bending	No Crack		2.5 Tons per diameter per kind
PP-02e	Concrete Works , supply, fabrication, delivery and erection	Reinforcing Steel	Reinforcing Steel	Tensile (Fy) 12mm dia. Bar and larger	413.7Mpa (60 ksi)	Structural Plan / Structural Specification	2.5 Tons per diameter per kind
				10mm dia bar And smaller	276 Mpa (40 ksi)		2.5 Tons per diameter per kind
				Bending	No Crack		

		Company Address :
Your Logo here!		Telephone : Fax : Email : info@ Website : www.

Inspection Request

		Date Prepared : 02/28/15	Template ID: MCT-Q-002
Project	Name of the Project	Revision : 000	Pages: 1 of 1
Project Location	Name of the location of the project	Reference No. : CM-PEC-D-0071-2015	

Contractor's/Seller's Company Name : Company A	Prepared by : Engineer X	Request No. 1

Mark applicable item below

Architectural		Structural	
Electrical		Mechanical	
Fire Protection Works		Others	

Description :

Received by(Inspector/Consultant) :	Inspection	
	Date:	Time:

Instruction / Comments :

Inspected/Commented by :	Forwarded/ Received by(Contractor/Seller):
Name and Signature: Role Company Date	Name and Signature of Seller/Contractor : Role Company Date

Your Logo here!

Company Address :
Telephone :
Fax :
Email : info@
Website : www.

Non Compliance Notice (NCN)

Date Prepared: 02/28/15	**Template ID:** MCT-Q-003		
Project	Name of the Project	**Revision :** 000	**Pages :** 1 of 1
Project Package	Name of the package / phase of the project		
Project Location	Name of the location of the project	**Reference No. :** CM-PEC-D-0072-2015	

Contractor's/Seller's Company Name	Contractor's/Seller's Name: In charge
Company A	Engineer X **Role :** Construction Manager
WBS ID:	**Scope / Work Package / Requirement:** Retaining Wall	**Reference :** (Drawing , Specification, Contract provision etc.)

No.	Description	Structure location /Particular	Current Status / Condition
1	Spacing of vertical Reinforcement	Retaining Wall at Grid line 1 and Grid line D-G	Installation of reinforcement steel bar

Issued by ;

_____
QC / Inspector

Noted by ;

_____
Project Manager

Received by ;

_____
Contractor / Seller

Date Received ;

_____

Punch Lists

Company Address :	
Telephone :	
Fax :	
Email : info@	
Website : www.	

Date Prepared : 02/28/15	Template ID: MCT-Q-003a
Revision : 000	Pages : 1 of 1
Reference No.: CM-PEC-D-0095-2015	

Project	Name of the Project
Project Package	Name of the package / phase of the project
Project Location	Name of the location of the project

Punch Lists

Item No.	Category	Deliverables	Area/Location	Description	Action	Date Issued	Date Closure Target	Date Closure Actual	Status	Remarks
1	Architectural	Ceiling	4th Floor, Room 401	Bottom of beam and slab – Shrinkage Cracks	Apply Brand Coat for all cracks				Still on progress	Ongoing application of coats, 80% done

Prepared by : **Inspector**

Noted by : **Consultant**

Company Address :
Telephone :
Fax :
Email : info@
Website : www.

Non Compliance Action Form

		Date Prepared: 02/28/15	Template ID: MCT-Q-004
Project	Name of the Project	Revision: 000	Pages: 1 of 1
Project Package	Name of the package / phase of the project		
Project Location	Name of the location of the project	Reference No. : CM-PEC-D-0073-2015	

Contractor's/Seller's Company Name	Contractor's/Seller's Name : In charge
Company A	Engineer X Role : Construction Manager
WBS ID : **Scope / Work Package / Requirement:** Retaining Wall	**Reference :** (Drawing , Specification , Contract provision etc.)

Corrective Action / Defect Repair

No.	Description	Action	Date of Completion	Inspector / Consultant's Comment
1	Spacing of vertical Reinforcement	Re work and corrected as per specified spacing		

Certified by ;

_____ / Date _____
Contractor's Quality Control

Verified by ;

_____ / Date _____
Contractor's Construction Manager

Received / Approved ;

_____ / Date _____
Inspector / Consultant

Noted by ;

_____ / Date _____
Project Manager

Performance Report

Company Address :	
Telephone :	
Fax :	
Email : info@	
Website : www.	

Date Prepared : 02/28/15	Template ID: MCT-COM-001
Revision : 000	Pages : 1 of 4
Reference No. :	CM-PEC-R-0009-2015

Project	Name of the Project
Project Package	Name of the package / phase of the project
Project Location	Name of the location of the project

PROGRESS – (Over all Work Accomplished)

You can plot the schedule in terms of bar chart with as planned vs. actual accomplishment.

WBS ID	Task Name	Duration	Start	Finish
1	Bored Piling Project	56 days?	Thu 4/25/13	Thu 7/11/13
1.1	General Requirements	47 days	Thu 4/25/13	Fri 6/28/13
1.1.1	Mobilization	13 days	Thu 4/25/13	Mon 5/13/13
1.1.2	Temporary Facilities Installation/erection	21 days	Thu 4/25/13	Thu 5/23/13
1.1.3	Material Handling and logistics	42 days	Thu 5/2/13	Fri 6/28/13
1.1.4	Demobilization	14 days	Mon 6/10/13	Thu 6/27/13
1.2	Earth Works	47 days	Thu 4/25/13	Fri 6/28/13
1.2.1	Trimming and clearing	43 days	Thu 4/25/13	Fri 6/28/13
1.2.2	Hauling of excavated and cleared materials	14 days	Tue 4/30/13	Fri 5/17/13
1.3	Load Bearing and Foundation Works	56 days	Thu 4/25/13	Thu 7/11/13
1.3.1	Steel casing fabrication	21 days	Thu 4/25/13	Thu 5/23/13
1.3.2	Bored Piling	35 days	Thu 5/2/13	Wed 6/19/13
1.3.3	Top of Pile cutting off	21 days	Wed 5/22/13	Wed 6/19/13
1.3.4	Fabrication of Rebar Case	35 days	Fri 5/24/13	Thu 7/11/13
1.3.5	Survey and Lay outing	35 days	Fri 5/24/13	Thu 7/11/13
1.4	Sign off	1 day?	Thu 7/11/13	Thu 7/11/13

This illustration, Lists is for illustrative only and not intended to portray any construction projects, scope, description or not representing any specific ways to organize, plan of any type of construction projects

www.ConstructionProjectManagementPro.com

	Company Address :
	Telephone :
	Fax :
	Email : info@
	Website : www.

Date Issue :	Pages :
02/28/15	2 of 4
Reference No. :	
CM-PEC-R-0009-2015	
Revision No. : 000	

Performance Report

STATUS – (stands on performance measurement baseline)

SCHEDULE

WBS ID	Work Package / Description	Planned %	Actual %	Variance	Status
1.1	General Requirements				
1.2	Mobilization / Demobilization				
1.3	Bored Pile	9.94	0.13	-9.81	19 Calendar days behind the schedule
1.4	Excavation/ Backfill				
1.5	Masonry				
1.6	Structural Steel				
1.6.1	Fabrication				
1.6.2	Erection				
1.7	Architectural				
1.8	Electrical				

This illustration, Lists is for illustrative only and not intended to portray any construction projects, scope, description or not representing any specific ways to organize, plan of any type of construction projects

www.ConstructionProjectManagementPro.com

Company Address :	
Telephone :	Fax :
Email : info@	
Website : www.	

Performance Report

Date Issue : 02/28/15	Pages : 3 of 4
Reference No. : CM-PEC-R-0009-2015	
Revision No. : 000	

COST :
(You can use also the plot of schedule with s-curve)

WBS ID	Work Package / Description	Planned %	Actual %	Variance	Status
1.1	General Requirements				
1.2	Mobilization / Demobilization				
1.3	Bored Pile				
1.4	Excavation/ Backfill				
1.5	Masonry				
1.6	Structural Steel				
1.6.1	Fabrication				
1.6.2	Erection				
1.7	Architectural				
1.8	Electrical				

This illustration, Lists is for illustrative only and not intended to portray any construction projects, scope, description or not representing any specific ways to organize, plan of any type of construction projects.

Performance Report

Date Issue: 02/28/15
Reference No.: CM-PEC-R-0009-2015
Revision No.: 000
Pages: 4 of 4

RISKS AND ISSUES — (stands on performance measurement baseline)

Risk	Status

Issue	Status

Risk Audit

Company Address :	
Telephone :	
Fax :	
Email : info@	
Website : www.	

Date Prepared : 02/28/15	Template ID: MCT-R-001
Revision : 000	Pages : 1 of 1
Reference No. : CM-PEC-D-0074-2015	

Project	Name of the Project
Project Package	Name of the package / phase of the project
Project Location	Name of the location of the project

No.	Risks				Response	
	Risks	Root Cause	Response Description	Remarks	Response improvements	Remarks

Risk Management Process

Process	Techniques	Process Improvements	Remarks

Company Address :
Telephone :
Fax :
Email : info@
Website : www.

Owner Supplied Materials (OSM)
Purchased and Delivered

Template ID:	MCT-P-001
Date Prepared :	02/28/15
Revision :	000
Pages :	1 of 1
Reference No.:	CM-PEC-D-0075-2015

Project	Name of the Project
Project Package	Name of the package / phase of the project
Project Location	Name of the location of the project

No.	Description	Manufacturer /Supplier	Date		Quantity		Unit	Contract Reference	Status
			Material Request	Delivery	Total	Delivered			

Owner Supplied Materials (OSM) - Consumption

Company Address :	
Telephone :	
Fax :	
Email : info@	
Website : www.	

Date Prepared : 02/28/15	Template ID: MCT-P-0002
Revision : 000	Pages : 1 of 1
Reference No. : CM-PEC-D-0076-2015	

Project	Name of the Project
Project Package	Name of the package / phase of the project
Project Location	Name of the location of the project

No.	Description	Seller / Supplier	Quantity		Consumption			Balance as of date	Status
			Total	Delivered	Previous	This Period	Total as of date		

Company Address:
Tel :0000000
Fax : 0000000
Email : info@
Website: www

Template ID: MCP-P-003
Revision: 000

Reference No. : CM-PEC-D-0077-2015

Date: February 28, 2015
Page 1 of 1

ORDER FOR PAYMENT

ATTENTION : Name in charge / Owners Representative for Accounts
Accounting/Administrative Division

PARTICULARS :
We hereby recommend the release of payment of **Supplier's Name, Supplier's Company Name ,** for the items/materials delivered on site and to be used for *specify the work package/ particular work.*

Project :		Date Prepared :	
Project Package :			
Project Location :			
ITEM	DESCRIPTION	AMOUNT	COMMENTS

Attachment;
- ✓ Delivery Receipts
- ✓ Invoice
- ✓ Others

Checked by:
_____
Site Personnel/ Inspector

Authorized by:
_____
Construction Manager/Project Manager

Approved by:
_____
Sponsor /Customer

www.ConstructionProjectManagementPro.com

Company Address:
Tel : 0000000
Fax : 0000000
Email : info@
Website: www

Template ID: MCT-P-004
Revision: 000

Reference No. : CM-PEC-D-0078-2015

BILLING INVOICE

SPONSOR / CUSTOMER DETAILS	Invoice Date :	Due Date : 30 days only
SOLD TO : Sponsor's / Customers' Company Name **ADDRESS :** Project Location **ATTENTION :** Sponsor's Name or Owner's Representative	Invoice No. :	Payment/Billing No.
	Project Code :	Contract No / Reference :

DESCRIPTION OF SERVICES		AMOUNT
Construction Management Services for the **Proposed Name of the Project** Located at **Location of the project/ Address**. Contract Reference No._____ Dated , **contract date** Contract Amount $_____ Services for the period of Month/ Day - Month/Day ,Year		
Prepared by : **Accountant**	**TOTAL SALE**	
Verified / Reviewed by: **Accounting Manager**	**VAT**	
Approved by: **Project Manager**	**AMOUNT DUE**	$

		Company Address :
Your Logo here!		Telephone :
		Fax :
		Email : info@
		Website : www. .

		Date Prepared:	Template ID:
Daily Time Sheet		02/28/15	MCT-P-005
Project	Name of the Project	**Revision :** 000	**Pages :** 1 of 1
Project Package	Name of the package / phase of the project		
Project Location	Name of the location of the project	**Reference No. :** CM-PEC-D-0079-2015	

Name	Regular Time (Reg)				Over Time (OT)		
	Time In	Signature	Time Out	Signature	Time In	Time Out	Signature

Certified by:

_____

Name / Signature
Site Administrative Officer

Noted by:

_____

Name/Signature
Project Manager

				Company Address :	
				Telephone :	
				Fax :	
				Email : info@	
				Website : www.	

Overtime Authorization Log

		Date Prepared: 02/28/15	Template ID: MCT-P-006
Project	Name of the Project	**Revision :** 000	**Pages :** 1 of 1
Project Package	Name of the package / phase of the project		
Project Location	Name of the location of the project	**Reference No. :** CM-PEC-D-0080-2015	

Name	Overtime Hours			Remarks
	Time In	Time Out	Total hours	

Authorized by:

_____
Name / Signature
Construction Manager

Noted by:

_____
Name/Signature
Project Manager

Request for Leave of Absence		Date Prepared: 02/28/15	Template ID: MCT-P-007
Project	Name of the Project	Revision: 000	Pages: 1 of 1
Project Package	Name of the package / phase of the project		
Project Location	Name of the location of the project	Reference No. : CM-PEC-D-0081-2015	

Company Address :
Telephone :
Fax :
Email : info@
Website : www.

Requested by :	Employee Number :	
	Division / Department :	
Nature of leave	( ) Sick	( ) Vacation
	( ) Emergency	( ) Maternity
	( ) Paternity	( ) Others
Period of leave	From:	To:
Days/Hours	Number of Days:	Equivalent Hours:

Reason for leave:

Approved by:

_____ _____
Site Staff Signature :   Construction Manager /Project Manager:

Human Resource Department (HRD) Verification	Type of leave	Available	Required	Balance
	sick			
	vacation			
	emergency			
	maternity			
	paternity			
	Others			

_____
Human Resource Officer

Your LOGO Here! – Contractor

Company Address:
Tel :0000000
Fax : 0000000
Email : info@
Website: www

Template ID: MCT-P-008
Revision: 000

Reference No. : CM-PEC-D-0082-2015

Date: February 28, 2015
Page 1 of 1

Attention	:	**PROJECT MANAGER'S NAME** Organization / Company Name
Project	:	**NAME OF THE PROJECT**
Project Package	:	PP 01 – General Civil and Architectural Works
Location	:	Project Location
Subject	:	**PROGRESS BILLING NUMBER 1** *(Generally, Construction Projects Have progress billings or a monthly billing based on their Accomplishments).*

Dear Sir,

Relative to the above project , we hereby request payment for the works completed covering periods from *month /day – month / day, year* with our accomplishment equivalent to **state percent** *(example 4.5 %)* amounting to **AMOUNT IN WORDS ($ - AMOUNT IN FIGURES).**

Attached here with are, site photo graphs, detailed computation of accomplishment, drawings or plans that shown portion of accomplished works.

If you have any clarifications, please advise us.

Respectfully Yours,

Signature:

_____
Name of the Seller / Contractor
    Role

_____
Organization or Company Name

Your LOGO Here! – Contractor

Company Address:
Tel :0000000
Fax : 0000000
Email : info@
Website: www

Template ID: MCT-P-009
Revision: 000

Reference No.: CM-PEC-D-0083-2015

Date: February 28, 2015

CONTRACTOR'S COMPANY NAME
Address

RELEASE OF LIENS

To All Whom it May Concern:

For valuable consideration, the undersigned has been hired as a CONTRACTOR by **PROJECT SPONSOR/CUSTOMER** to render all labor, materials, equipment and services for **Project Name, Project Package,** located at Project location and address.

NOW THEREFORE, this **2ndday of May**(Month) ,**2013** (year), the undersigned hereby releases the property of the **PROJECT SPONSOR/CUSTOMER**, located at Project location and address,

From any liability from lien for all labor, materials, equipment and services rendered and or delivered for that said property.

CONTRACTOR'S COMPANY NAME
Contractor

_____
Authorized Representative Signature

_____**Name**_____
Role

"This Release of Lien Template/Form is for illustrative only , Consult your legal adviser or attorney for further details."

Your LOGO Here!	Company Address: Tel:0000000 Fax: 0000000 Email: info@ Website: www
Reference No. : CM-PEC-D-0084-2015	Template ID: MCT-P-010 Revision: 000 Date: February 28, 2015

CONTRACTOR'S COMPANY NAME
Address

CONTRACTOR'S RELEASE OF LIENS UPON FINAL PAYMENT

To All Whom it May Concern:

For valuable consideration, the undersigned has been hired as a CONTRACTOR by **PROJECT SPONSOR/CUSTOMER** to render all labor, materials, equipment and services for **Project Name, Project Package**, located at Project location and address .

NOW THEREFORE, this **2nd day of May**(Month) ,**2013** (year) , upon the receipt of the undersigned for and in consideration of the final amount from **PROJECT SPONSOR/CUSTOMER** the sum of $_____ as final payment, does hereby waive and release any lien rights the undersigned has on the property of **PROJECT SPONSOR/CUSTOMER NAME**, located at Project location.

CONTRACTOR'S COMPANY NAME
Contractor

_____
Authorized Representative Signature

_____
Name
Role

"This Release of Lien Template/Form is for illustrative only, Consult your legal adviser or attorney for further details."

Your LOGO Here!

Company Address:
Tel:0000000
Fax: 0000000
Email: info@
Website: www

Template ID: MCT-P-011
Revision: 000

Reference No. : CM-PEC-D-0085-2015

Date: February 28, 2015
Page 1 of 1

SURETY COMPANY NAME
Address

CONSENT OF SURETY UPON FINAL PAYMENT

Project : P.E.C. School Building Project – (Project Name)
Project Package :
Project Location :

In consideration of the contract made between the Contractor and the Sponsor/Customer, the surety company, **SURETY COMPANY NAME, ADDRESS, BRANCH)** on the Payment Bond of the Contractor, **CONTRACTOR'S NAME, ADDRESS,** is hereby approves the issuance of final payment, said final payment does not relieve the surety company to it's obligation to the Sponsor / Customer, **NAME OF THE SPONSOR/CUSTOMER, ADDRESS**

IN WITNESS WHERE OF, We have set our hands this **21stday** of **May** (month), **2013**(year).

SURETY COMPANY NAME

By:

_____**Name**_____
Branch Manager

_____**Name**_____
Senior Manager

"This Consent Surety Template/Form is for illustrative only, Consult a surety company near you for details"

Reference No. : CM-PEC-L-0004-2015

Date: February 28, 2015
Page 1 of 1

SPONSOR'S / CUSTOMER'S COMPANY NAME

Address

Attention	:	**PROJECT SPONSOR/ CUSTOMER'S NAME**
Project	:	**NAME OF THE PROJECT**
Project Package	:	PP 01 – General Civil and Architectural Works
Location	:	Project Location
Subject	:	**RECOMMENDATION FOR PROGRESS BILLING No. 01**
Contractor /Seller	:	**CONTRACTOR / SELLER'S COMPANY NAME**

Gentlemen :

We recommend herewith, for your approval, the request for first progress payment of **Contractor's / Seller's Company Name** for the *Project Package* of the *Project Name* . We have evaluated the request and hereby recommend payment in the total amount of **USD:*AMOUNT IN WORDS* ($ - Amount if figures**)representing *percentage %(example 4.5%)* of the work accomplishment.

Attached are the billing request of **CONTRACTOR /SELLER'S COMPANY NAME** dated *state the date of request for payment letter of contractor / seller*, the evaluated accomplishment report as of date where the evaluation report was made.

Trusting you will find the above in order.

Very truly yours   Noted by:

_____   _____
Construction Manager   Project Manager

Company Address:
Tel :0000000
Fax : 0000000
Email : info@
Website: www

Template ID: MCT-P-013
Revision: 000

Reference No.: CM-PEC-D-0086-2015

Date: February 28, 2015

Page 1 of 1

PAYMENT DETAILS AND COMPUTATION

This is to certify that in accordance with the terms and conditions of the contract executed on *Date and Year of the contract (Day 1)* by and between (OWNER) *Name of the Sponsor* and (CONTRACTOR) **the company name of the Contractor/ Seller** for the *Project Package Name* of the **PROPOSED NAME OF THE PROJECT**, located at Project location.

There will be due and payable from the Owner to the Contractor the sum of USD : **AMOUNT IN WORDS**		H
1. Contract Amount	A	
2. Additions / Deductions (impact from changes)		
Change Order No. 01	B	
3. Contract Amount to date(with effect of changes)	C	
4. Value of Work completed to date, 10.498%		D
5. Less:		
a. Retention 10%	10% of Contract Amount , verify contract condition	
b. Repayment of downpayment,	pro rate the value of down payment with total dproject duration	
Sub-Total (Deductions)		E
6. Total amount due to Contractor		F
7. Less: Previous Progress Payment	G	
8. Amount due to Contract Now		H
9. Contract Balance including Retention		I
10. Summary of Payments	J+G+H	
10.1 Advance Payment	J	
10.2 First Progress Billing	G	
10.3 Second Progress Billing (This Billing)	H	

EVALUATED PERCENTAGE OF WORK COMPLETED TO DATE: ___ %

CERTIFIED CORRECT BY: NOTED ;

NAME   **PROJECT MANAGER'S NAME**
Construction Manager   Project Manager

Daily Time Sheet - Seller

Company Address :	
Telephone :	
Fax :	
Email : info@	
Website : www.	

Project	Name of the Project	**Date Prepared:** 02/28/15	**Template ID:** MCT-P-014
Project Package	Name of the package / phase of the project	**Revision :** 000	**Pages :** 1 of 1
Project Location	Name of the location of the project	**Reference No. :** CM-PEC-D-087-2015	

Name	Regular Time (Reg)				Over Time (OT)		
	Time In	Signature	Time Out	Signature	Time In	Time Out	Signature

Certified by:

_____

Name / Signature
Site Administrative Officer

Noted by:

_____

Name/Signature
Project Manager

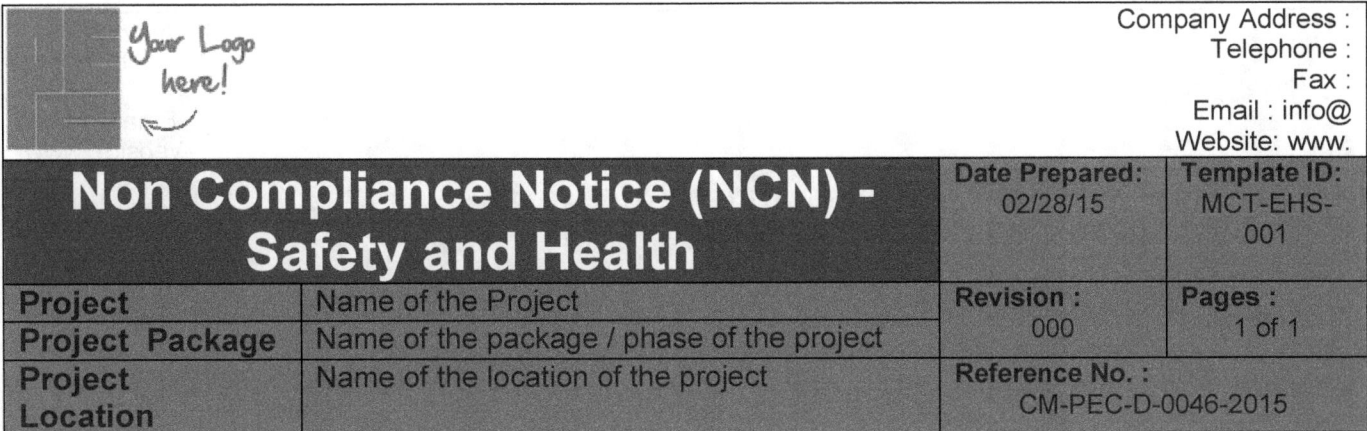

Non Compliance Notice (NCN) - Safety and Health		Date Prepared: 02/28/15	Template ID: MCT-EHS-001
Project	Name of the Project	Revision: 000	Pages: 1 of 1
Project Package	Name of the package / phase of the project		
Project Location	Name of the location of the project	Reference No.: CM-PEC-D-0046-2015	

Contractor's/Seller's Company Name Company A	Contractor's/Seller's Name : In charge Engineer X Role : Construction Manager

NOTICE NO. 001	Scope / Work Package / Requirement:	Reference :(Contract provision and others)

Description	Specific location (Working Area)	Due Date
Safety:		
Hard Hat		
Safety shoes		
Rain Gear		
Gloves		
Goggles/Safety Glasses		
Safety Full Body Harness		
Scaffolding		
Fire Extinguisher		
Electrical connection		
Ladder		
Maintenance Tool		
Safe Mask		
Face Shield		
Work permit		
Safety Meeting		
Others;		
Health:		
Health facility		
First Aid		
Others;		

Issued by ;
_____
Safety Officer/ Health Officer
Noted by ;
_____
Project Manager

Received by ;
_____
Contractor / Seller

Date Received ;
_____

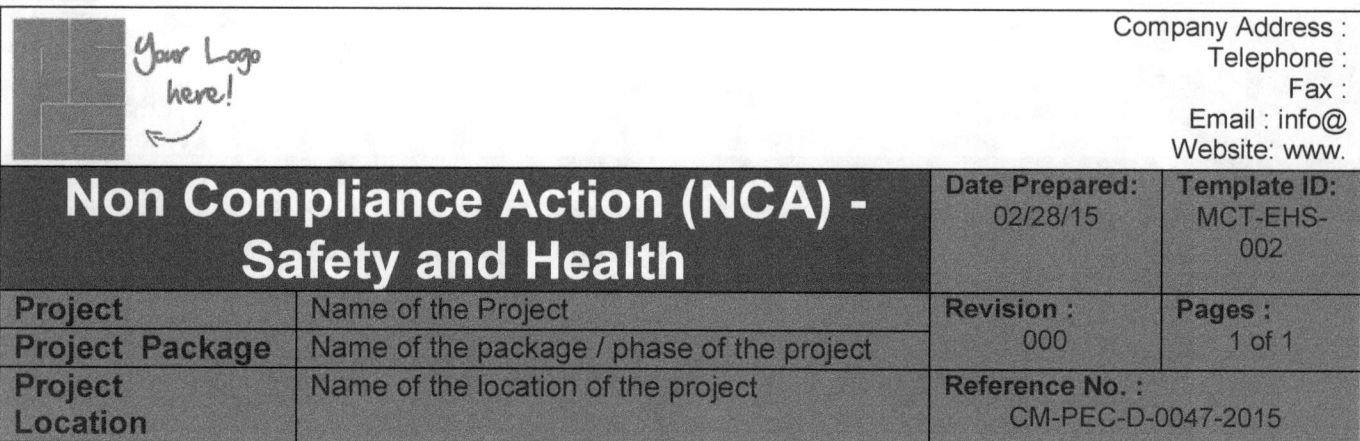

Non Compliance Action (NCA) - Safety and Health		Date Prepared: 02/28/15	Template ID: MCT-EHS-002
Project	Name of the Project	Revision: 000	Pages: 1 of 1
Project Package	Name of the package / phase of the project		
Project Location	Name of the location of the project	Reference No.: CM-PEC-D-0047-2015	

Contractor's/Seller's Company Name Company A		Contractor's/Seller's Name : In charge Engineer X Role: Construction Manager	
NOTICE NO. 001	Scope / Work Package / Requirement:	Reference :(Contract provision and others)	

Corrective / Preventive Action

No.	Description	Action	Date of Completion	Inspector / Consultant's Comment
1				

Certified by;

_____ / Date _____
Contractor's Quality Control

Verified by;

_____ / Date _____
Contractor's Construction Manager

Received / Approved;

_____ / Date _____
Inspector / Consultant

Noted by;

_____ / Date _____
Project Manager

Step 5
Closing

www.ConstructionProjectManagementPro.com

STEP 5 : CLOSING

Closing is the fifth and last step in managing construction projects with the purpose of formally completing the contractual obligation and the project or project phase ,This assures that all processes from established project management plan are completed to close the project phase or project .

Flow Chart - Project Management process is not always sequential or performed in identical sequence.

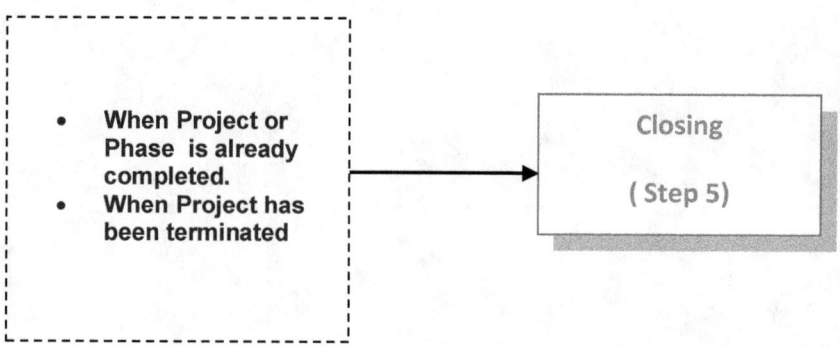

Mapping:

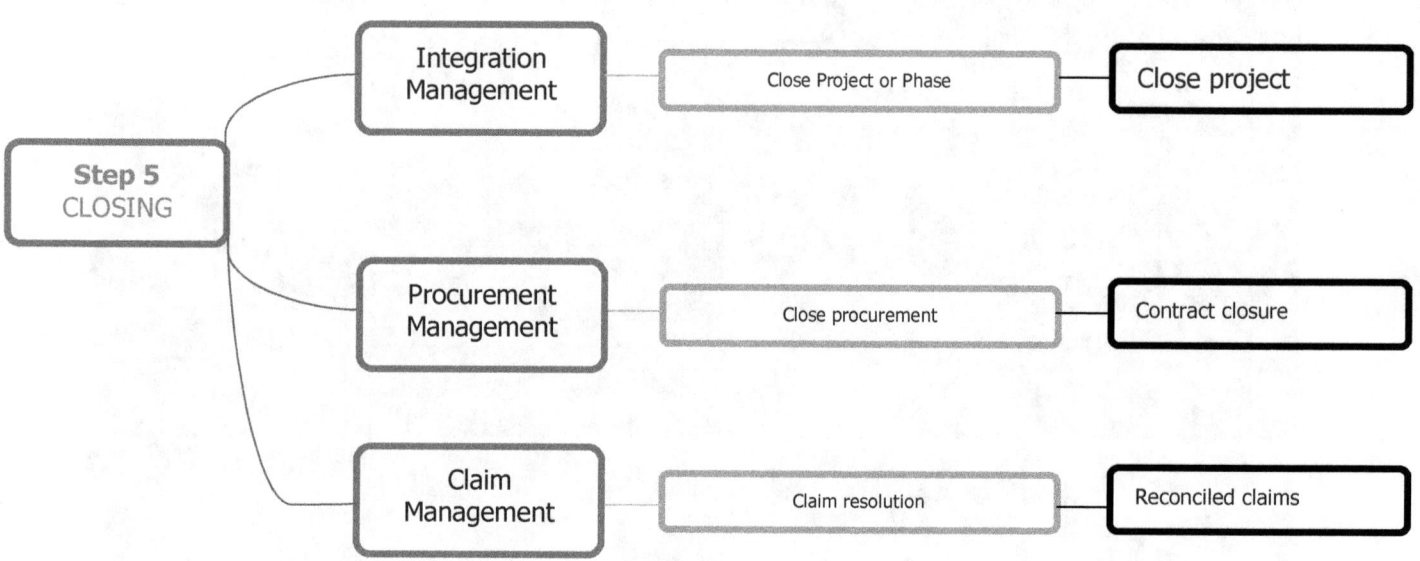

CLOSING PROJECT OR PHASE

The closing project or phase involves checking, verifying, and finalizing the project activities based on established project management processes. All projects must have a formal close out process, whether it is terminated or accepted. This document is being created during the fifth step of managing construction projects.

Template : Close project

CONTRACT CLOSURE

Contract closure refers to the phase where contract must be closed, including all the information gathered like contract changes, payment records, and performance on the scope, quality, schedule, and cost. This information should be verified and submitted as they will be used in evaluating contractors for future contacts. This document is to be created during the fifth of managing construction projects.

Template : Contract Closure , Reconciled Claims

	Close Project	Date Prepared: 02/28/15	Template ID: CT-I-001
Project	Name of the Project	**Revision:** 000	**Pages:** 1 of 2
Project Package	Name of the package / phase of the project		
Project Location	Name of the location of the project	**Reference No.:** CM-PEC-D-0088-2015	

Company Address :
Telephone :
Fax :
Email : info@
Website : www.

PROJECT DESCRIPTION:

Description	Cost		Schedule	
	Planned	Actual	Planned	Completed

CONTRACT

			DESCRIPTION				
Category	Contract ID	Number of Pages	Date		Recipient		
			Prepared	Received by Vendor	Name	Role	Organization

	REVISION	
ID	Description	Approval Date

DELIVERABLES:

ID	Category	Deliverables	Acceptance Detail		Comments
			Date	Reference	

	Close Project	Company Address : Telephone : Fax : Email : info@ Website : www.
		Date Issue : 02/28/15 — Pages : 2 of 2 Reference No.: CM-PEC-D-0088-2015 Revision No. : 000

PROJECT PERFORMANCE ANALYSIS

Technical	
Scope	
Schedule	
Cost	
Quality	
Requirements	
Others	
Decision making	
Issues	
Communication	

Approvals:

Pier Engineering and Consultants   **P.E.C School**

By: By:

_____   _____   _____

Pier John   **Alexander Thomas**   **Richard Ramos**
Project Manager   Vice President   President

		Company Address: Telephone: Fax: Email: info@ Website: www.	
Contract Closure		Date Prepared: 02/28/15	Template ID: CT-P-001
Project	Name of the Project	Revision: 000	Pages: 1 of 3
Project Package	Name of the package / phase of the project		
Project Location	Name of the location of the project	Reference No.: CM-PEC-D-0089-2015	

SUPPLIER / VENDOR PERFORMANCE ANALYSIS

Technical	
Scope	
Schedule	
Cost	
Quality	
Requirements	
Others	
Decision making	
Issues	
Communication	

Contract Closure

Date Issue :	02/28/15
Reference No. :	CM-PEC-D-0089-2015
Revision No. :	000

FOR IMPROVEMENT

Technical	
Category	**Corrective Action**
Scope	
Schedule	
Cost	
Quality	
Requirements	

Others	
Category	**Corrective Action**
Decision making	
Issues	
Communication	

		Company Address :
Your Logo here!		Telephone :
		Fax :
		Email : info@
		Website : www.
Contract Closure	Date Issue : 02/28/15	Pages : 3 of 3
	Reference No. : CM-PEC-D-0089-2015	
	Revision No. : 000	

CONTRACT DESCRIPTION:

Category	Contract ID	Number of Pages	Date		Recipient		
			Prepared	Received by Vendor	Name	Role	Organization

CONTRACT CHANGES:

ID	Description	Approval Date

CONTRACT PAYMENTS:

Payment Number	Date		Amount		Date (Payment Date)
	Forwarded by Vendor	Evaluated / Checked	Claimed / Bill Amount	Approved (Amount to be received)	

Approval:

Pier Engineering and Consultants   **P.E.C School**

By: By:

_____   _____   _____

Pier John   **Alexander Thomas**   **Richard Ramos**
Project Manager   Vice President   President

Reconciled Claim

Company Address :			Date Prepared :	09/01/13
Telephone :			Template ID:	CT-CL-001
Fax :			Revision :	000
Email : info@			Pages :	1 of 1
Website : www.			Reference No. :	CM-PEC-D-0095-2013

Project	Name of the Project
Project Package	Name of the package / phase of the project
Project Location	Name of the location of the project

Claim No	Category / WBS	Claim and Description	Consultant's Estimate	Contractor's Estimate	Reconciled Amount / Quantity	Comments / Justification
1		Force Majeure, 10 Days Time Extension	9 days	10 days	9 days	
2						
3						
4						
5						
6						
7						
8						
9						

Contractor :   Project Owner   Consultant:

Construction Management: Step-by-Step Templates   Training

Manage Templates and Documents the Easy Way.
You can easily manage your entire project templates and documents as easy as shown below!.

Let's have the Step 1 : Initiating (Initiate your project)

The first part of the document is the step *templates (initiating templates)* , as you can see below, this is your controlling documents, where in you can log templates and documents with this specific step (step 1) . This is your document that is generally kept in your main office or company management files, provide a copy in your site files that has to be maintained by your document controller, some companies treated this as a confidential document.

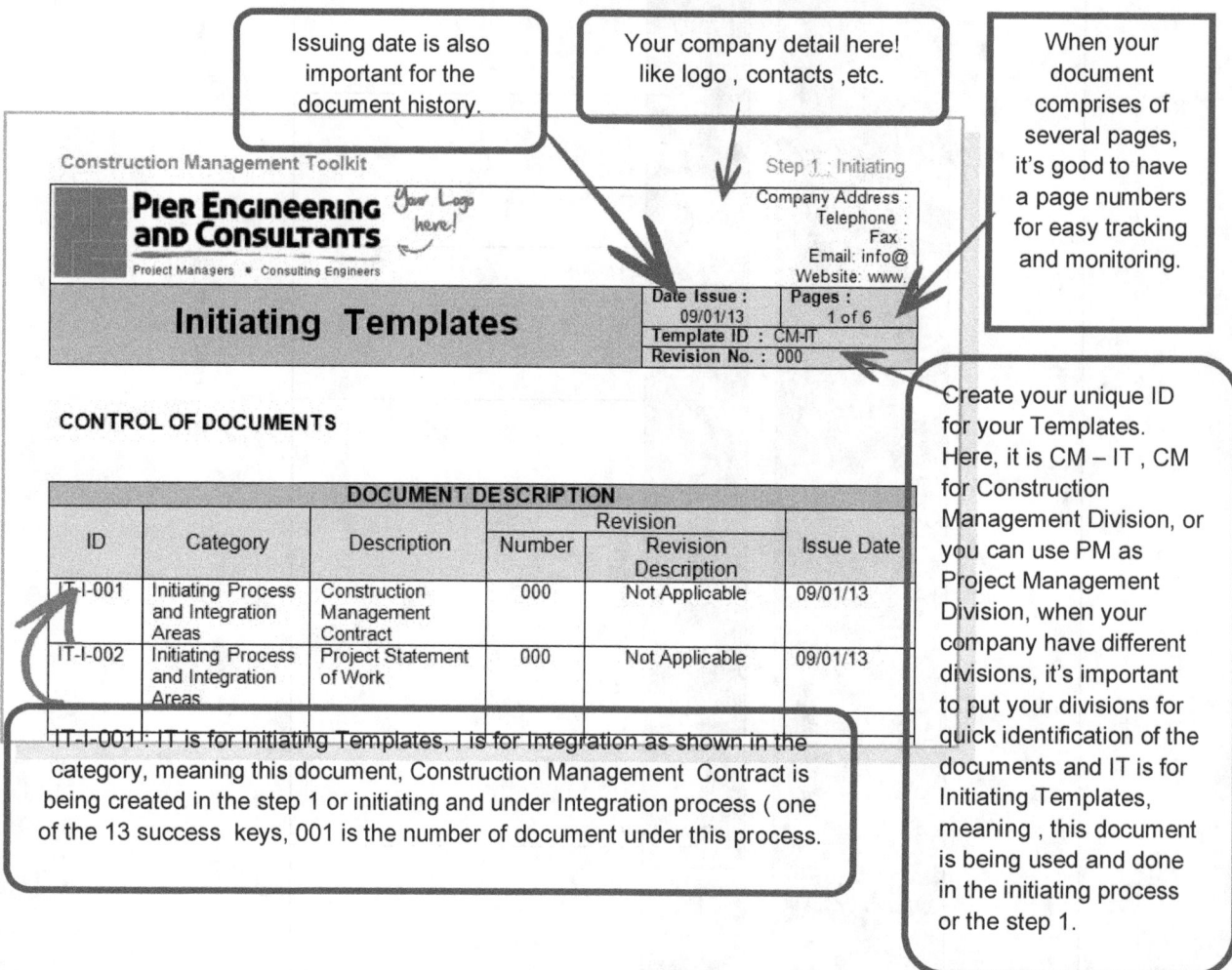

Note : You can tailor or edit these initials, codes and identification based on your choice or company preference, what is important here is that, you can easily distinguish what type of document, where to use, what steps and success keys you're in.

www.ConstructionProjectManagementPro.com

Construction Management: Step-by-Step Templates Training

Here are the templates and documents for Step 1 : Initiating (Initiate your project)

> **IT-I-001** : **IT** means this construction management contract is being created and provided in the step 1 or initiating process, **I** is for Integration Management, meaning this document is for step 1 under Integration management and a first or number one (**001**) in this process.

4.0 Templates

Number	Form	Template ID	Number of pages
1	Construction Management Contract	IT-I-001	1
2	Project Statement of Work	IT-I-002	2
3	Project Charter	IT-I-003	5
4	Stakeholder Registry	IT-COM-001	1

www.ConstructionProjectManagementPro.com

> **IT-COM-001** : **IT** means this stakeholder registry is being created and provided in the step 1 or initiating process, **COM** is for Communication Management, meaning this document is for step 1 under communication management and a first or number one (**001**) in this process.

IT - Initiating Process or our Step 1
I - Integration Management (one of the 13 success keys)
COM – Communication Management (one of the 13 success keys)
001 and others – Document number for every success keys.

Note : You can tailor or edit these initials, codes and identification based on your choice or company preference, what is important here is that, you can easily distinguish what type of document, where to use, what steps and success keys you're in.

FOR LETTERS, DOCUMENTS AND REPORTS

> CM - for Division or Construction Management Division (CM Division), you can use PM as Project Management Division
>
> PEC - Your company name
>
> D – This is a document
>
> 0001 – Document number under category of document.
>
> 2013 – The year created this document.

> **IT-I-003 :** IT means this project charter is being created and provided in the step 1 or initiating process, I is for Integration Management, meaning this document is for step 1 under Integration management and the third or number three (003) in the document for Integration management.

Construction Management Toolkit

Reference No. : CM-PEC-D-0001-2013

Step 1 : Initiating

Template ID : IT-I-003
Revision : 000

Date : September 01 , 2013
Page 1 of 5

PROJECT CHARTER

1.0   PROJECT DESCRIPTION
This shows a descriptive summary of the project.

The Management of P.E.C. School have noticed the increase of college students every school year and decided to have an additional school building to accommodate the increasing number of enrollees

---

D – This is a Document
L - This is a Letter
R – This is a Report

Note : You can tailor or edit these initials, codes and identification based on your choice or company preference, what is important here is that, you can easily distinguish what type of document, where to use, what steps and success keys you're in.

190

www.ConstructionProjectManagementPro.com

Construction Management: Step-by-Step Templates   Training

FOR TEMPLATES

As always mentioned in this publication, all Templates and Project Documents has been properly coded and arranged in sequence, see illustration below how it works!.

Project Name and Location - It's perfect to start your Forms, Templates with the name of the project, it's the title of what project you are working for and this template referring for. Location is important also as reference and identification, some projects you can easily remind through location.

Package – This is useful to identify what specifically you are working with. Let's say the project is a large Mall and your specific project package is only the foundation works (bored piling works etc), as a project manager of foundation project, you can easily separate your specific project package with the entire

TEMPLATE ID : IT-COM-001 -

> **IT** - means this Stakeholder Registry is being created and provided in the step 1 or initiating process.
> **COM** - is for Communication Management, meaning this document is for step 1 under Communication Management.
> **001** - Is the first or the number one in the document or file for communication management.

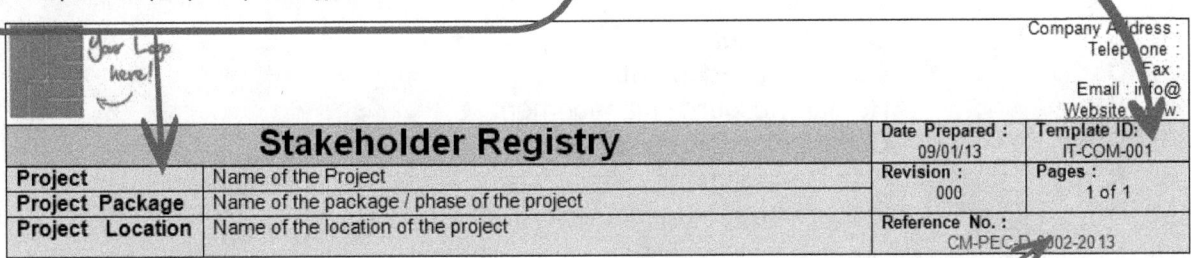

Reference No. – This is your reference number for every specific document or a template.

CM-PEC-D-0002-2013 :

CM : Construction Management Division, some companies have different divisions, like CM, Design Division or you can use **PM** as Project Management Division etc.

PEC is for your company initial or name

D is for a document, this template is a kind of Document.

0002 – Is the number of document, meaning the stakeholder template is the second template for a Document and not a Letter or Report.

2013 – The year when did you created this type of document.

USEFUL ABBREVIATION AS TAG OR A CODE TO EASILY MANAGE YOUR CONSTRUCTION DOCUMENTS.

- **5 Easy Steps**

    1. **IT**   : Initiating Templates
    2. **PT**   : Planning Templates
    3. **ET**   : Executing Templates
    4. **MCT** : Monitoring and Controlling Templates
    5. **CT**   : Closing Templates

- **13 Success Keys**

    1. **I**   : Integration Management
    2. **SC**   : Scope Management
    3. **T**   : Time Management
    4. **C**   : Cost Management
    5. **Q**   : Quality Management
    6. **HR**   : Human Resource Management
    7. **COM** : Communication Management
    8. **R**   : Risk Management
    9. **P**   : Procurement Management

Note : You can tailor or edit these initials, codes and identification based on your choice or company preference, what is important here is that, you can easily distinguish what type of document, where to use, what steps and success keys you're in.

www.ConstructionProjectManagementPro.com

Get Your Free Templates, Flow charts, Articles, Examples, videos for construction projects and Learn the ins and outs of Project Management for Construction projects.

www.ConstructionManagementToolkit.com

The Complete and Step-by-Step Guide to Construction Management, You will discover all the processes involved in project management for construction projects, from 5 Easy Steps, Templates, Form, Flowcharts, Example, Tips and Tricks To 13 Success Keys, all are arranged in sequence and all templates and documents are fully customizable.

www.PierEngineeringandConsultants.com

The Company for Project Managers and Consulting Engineers.

www.Facebook.com/constructionprojectmanagementpro

Like Us on FACEBOOK to get the latest update, free tools announcement and more..

As a purchaser of this book you will also receive an electronic file through email (*just send us email*) of all the templates and documents for you to personalize and customize in your own preference and liking.

As promised shown on our website…

YOUR FREE CONSTRUCTION MANAGEMENT TOOLS
($ 197.00 Value)

This Includes:

- ✓ Legal Document : Project Management Services Contract For Construction

- ✓ Bid Conference : Pre-bid Conference Presentation Documents

- ✓ WBS : An 8th Floor Building Project , You'll learn all the components of a Vertical building from work package to activity level, you can use this in all of your vertical projects as your guide, checklists and more.

> As a purchaser of this book you will also receive an electronic file through email (*just send us email*) of all the bonuses, templates and documents for you to personalize and customize in your own preference and liking.